Phase Noise and Frequency Stability in Oscillators

Presenting a comprehensive account of oscillator phase noise and frequency stability, this practical text is both mathematically rigorous and accessible. An in-depth treatment of the noise mechanism is given, describing the oscillator as a physical system, and showing that simple general laws govern the stability of a large variety of oscillators differing in technology and frequency range. Inevitably, special attention is given to amplifiers, resonators, delay lines, feedback, and flicker (1/f) noise. The reverse engineering of oscillators based on phase-noise spectra is also covered, and end-of-chapter exercises are given. Uniquely, numerous practical examples are presented, including case studies taken from laboratory prototypes and commercial oscillators, which allow the oscillator internal design to be understood by analyzing its phase-noise spectrum. Based on tutorials given by the author at the Jet Propulsion Laboratory, international IEEE meetings, and in industry, this is a useful reference for academic researchers, industry practitioners, and graduate students in RF engineering and communications engineering.

Additional materials are available via www.cambridge.org/rubiola.

Enrico Rubiola is a Senior Scientist at the CNRS FEMTO-ST Institute and a Professor at the Université de Franche Comté. With previous positions as a Professor at the Université Henri Poincaré, Nancy, and in Italy at the University Parma and the Politecnico di Torino, he has also consulted at the NASA/Caltech Jet Propulsion Laboratory. His research interests include low-noise oscillators, phase/frequency-noise metrology, frequency synthesis, atomic frequency standards, radio-navigation systems, precision electronics from dc to microwaves, optics and gravitation.

T0207130

The Cambridge RF and Microwave Engineering Series

Series Editor
Steve C. Cripps

Peter Aaen, Jaime Plá and John Wood, *Modeling and Characterization of RF and Microwave Power FETs*
Enrico Rubiola, *Phase Noise and Frequency Stability in Oscillators*
Dominique Schreurs, Máirtín O'Droma, Anthony A. Goacher and Michael Gadringer, *RF Amplifier Behavioral Modeling*
Fan Yang and Yahya Rahmat-Samii, *Electromagnetic Band Gap Structures in Antenna Engineering*

Forthcoming:

Sorin Voinigescu and Timothy Dickson, *High-Frequency Integrated Circuits*
Debabani Choudhury, *Millimeter Waves for Commercial Applications*
J. Stephenson Kenney, *RF Power Amplifier Design and Linearization*
David B. Leeson, *Microwave Systems and Engineering*
Stepan Lucyszyn, *Advanced RF MEMS*
Earl McCune, *Practical Digital Wireless Communications Signals*
Allen Podell and Sudipto Chakraborty, *Practical Radio Design Techniques*
Patrick Roblin, *Nonlinear RF Circuits and the Large-Signal Network Analyzer*
Dominique Schreurs, *Microwave Techniques for Microelectronics*
John L. B. Walker, *Handbook of RF and Microwave Solid-State Power Amplifiers*

Phase Noise and Frequency Stability in Oscillators

ENRICO RUBIOLA
Professor of Electronics
FEMTO-ST Institute
CNRS and Université de Franche Comté
Besançon, France

CAMBRIDGE UNIVERSITY PRESS
Cambridge, New York, Melbourne, Madrid, Cape Town, Singapore,
São Paulo, Delhi, Dubai, Tokyo

Cambridge University Press
The Edinburgh Building, Cambridge CB2 8RU, UK

Published in the United States of America by Cambridge University Press, New York

www.cambridge.org
Information on this title: www.cambridge.org/9780521153287

© Cambridge University Press 2009

First published 2009
This digitally printed version 2010

A catalogue record for this publication is available from the British Library

ISBN 978-0-521-88677-2 Hardback
ISBN 978-0-521-15328-7 Paperback

Contents

Foreword by Lute Maleki

Given the ubiquity of periodic phenomena in nature, it is not surprising that oscillators play such a fundamental role in sciences and technology. In physics, oscillators are the basis for the understanding of a wide range of concepts spanning field theory and linear and nonlinear dynamics. In technology, oscillators are the source of operation in every communications system, in sensors and in radar, to name a few. As man's study of nature's laws and human-made phenomena expands, oscillators have found applications in new realms.

Oscillators and their interaction with each other, usually as phase locking, and with the environment, as manifested by a change in their operational parameters, form the basis of our understanding of a myriad phenomena in biology, chemistry, and even sociology and climatology. It is very difficult to account for every application in which the oscillator plays a role, either as an element that supports understanding or insight or an entity that allows a given application.

In all these fields, what is important is to understand how the physical parameters of the oscillator, i.e. its phase, frequency, and amplitude, are affected, either by the properties of its internal components or by interaction with the environment in which the oscillator resides. The study of oscillator noise is fundamental to understanding all phenomena in which the oscillator model is used in optimization of the performance of systems requiring an oscillator.

Simply stated, noise is the unwanted part of the oscillator signal and is unavoidable in practical systems. Beyond the influence of the environment, and the non-ideality of the physical elements that comprise the oscillator, the fundamental quantum nature of electrons and photons sets the limit to what may be achieved in the spectral purity of the generated signal. This sets the fundamental limit to the best performance that a practical oscillator can produce, and it is remarkable that advanced oscillators can reach it.

The practitioners who strive to advance the field of oscillators in time-and-frequency applications cannot be content with knowledge of physics alone or engineering alone. The reason is that oscillators and clocks, whether of the common variety or the advanced type, are complex "systems" that interact with their environment, sometimes in ways that are not readily obvious or that are highly nonlinear. Thus the physicist is needed to identify the underlying phenomenon and the parameters affecting performance, and the engineer is needed to devise the most effective and practical approach to deal with them. The present monograph by Professor Enrico Rubiola is unique in the extent to which it satisfies both the physicist and the engineer. It also serves the need to understand both the fundamentals and the practice of phase-noise metrology, a required tool in dealing with noise in oscillators.

Rubiola's approach to the treatment of noise in this book is based on the input–output transfer functions. While other approaches lead to some of the same results, this treatment allows the introduction of a mathematical rigor that is easily tractable by anyone with an introductory knowledge of Fourier and Laplace transforms. In particular, Rubiola uses this approach to obtain a derivation, from first principles, of the Leeson formula. This formula has been used in the engineering literature for the noise analysis of the RF oscillator since its introduction by Leeson in 1966. Leeson evidently arrived at it without realizing that it was known earlier in the physics literature in a different form as the Schawlow–Townes linewidth for the laser oscillator. While a number of other approaches based on linear and nonlinear models exist for analyzing noise in an oscillator, the Leeson formula remains particularly useful for modeling the noise in high-performance oscillators. Given its relation to the Schawlow–Townes formula, it is not surprising that the Leeson model is so useful for analyzing the noise in the optoelectronic oscillator, a newcomer to the realm of high-performance microwave and millimeter-wave oscillators, which are also treated in this book.

Starting in the Spring of 2004, Professor Rubiola began a series of limited-time tenures in the Quantum Sciences and Technologies group at the Jet Propulsion Laboratory. Evidently, this can be regarded as the time when the initial seed for this book was conceived. During these visits, Rubiola was to help architect a system for the measurement of the noise of a high-performance microwave oscillator, with the same experimental care that he had previously applied and published for the RF oscillators. Characteristically, Rubiola had to know all the details about the oscillator, its principle of operation, and the sources of noise in its every component. It was only then that he could implement the improvement needed on the existing measurement system, which was based on the use of a long fiber delay in a homodyne setup.

Since Rubiola is an avid admirer of the Leeson model, he was interested in applying it to the optoelectronic oscillator, as well. In doing so, he developed both an approach for analyzing the performance of a delay-line oscillator and a scheme based on Laplace transforms to derive the Leeson formula, advancing the original, heuristic, approach. These two treatments, together with the range of other topics covered, should make this unique book extremely useful and attractive to both the novice and experienced practitioners of the field.

It is delightful to see that in writing the monograph, Enrico Rubiola has so openly bared his professional persona. He pursues the subject with a blatant passion, and he is characteristically not satisfied with "dumbing down," a concept at odds with mathematical rigor. Instead, he provides visuals, charts, and tables to make his treatment accessible. He also shows his commensurate tendencies as an engineer by providing numerical examples and details of the principles behind instruments used for noise metrology. He balances this with the physicist in him that looks behind the obvious for the fundamental causation. All this is enhanced with his mathematical skill, of which he always insists, with characteristic modesty, he wished to have more. Other ingredients, missing in the book, that define Enrico Rubiola are his knowledge of ancient languages and history. But these could not inform further such a comprehensive and extremely useful book on the subject of oscillator noise.

Lute Maleki
NASA/Caltech Jet Propulsion Laboratory
and OEwaves, Inc.,
February 2008

Foreword by David Leeson

Permit me to place Enrico Rubiola's excellent book *Phase Noise and Frequency Stability in Oscillators* in context with the history of the subject over the past five decades, going back to the beginnings of my own professional interest in oscillator frequency stability.

Oscillator instabilities are a fundamental concern for systems tasked with keeping and distributing precision time or frequency. Also, oscillator phase noise limits the demodulated signal-to-noise ratio in communication systems that rely on phase modulation, such as microwave relay systems, including satellite and deep-space links. Comparably important are the dynamic range limits in multisignal systems resulting from the masking of small signals of interest by oscillator phase noise on adjacent large signals. For example, Doppler radar targets are masked by ground clutter noise.

These infrastructure systems have been well served by what might now be termed the classical theory and measurement of oscillator noise, of which this volume is a comprehensive and up-to-date tutorial. Rubiola also exposes a number of significant concepts that have escaped prior widespread notice.

My early interest in oscillator noise came as solid-state signal sources began to be applied to the radars that had been under development since the days of the MIT Radiation Laboratory. I was initiated into the phase-noise requirements of airborne Doppler radar and the underlying arts of crystal oscillators, power amplifiers, and nonlinear-reactance frequency multipliers.

In 1964 an IEEE committee was formed to prepare a standard on frequency stability. Thanks to a supportive mentor, W. K. Saunders, I became a member of that group, which included leaders such as J. A. Barnes and L. S. Cutler. It was noted that the independent use of frequency-domain and time-domain definitions stood in the way of the development of a common standard. To promote focused interchange the group sponsored the November 1964 NASA/IEEE Conference on Short Term Frequency Stability and edited the February 1966 *Special Issue on Frequency Stability* of the *Proceedings of the IEEE*.

The context of that time included the appreciation that self-limiting oscillators and many systems (FM receivers with limiters, for example) are nonlinear in that they limit amplitude variations (AM noise); hence the focus on phase noise. The modest frequency limits of semiconductor devices of that period dictated the common usage of nonlinear-reactance frequency multipliers, which multiply phase noise to the point where it dominates the output noise spectrum. These typical circuit conditions were second nature then to the "short-term stability community" but might not come so readily to mind today.

The first step of the program to craft a standard that would define frequency stability was to understand and meld the frequency- and time-domain descriptions of phase instability to a degree that was predictive and permitted analysis and optimization. By the time the subcommittee edited the *Proc. IEEE* special issue, the wide exchange of viewpoints and concepts made it possible to synthesize concise summaries of the work in both domains, of which my own model was one.

The committee published its "Characterization of frequency stability" in *IEEE Trans. Instrum. Meas.*, May 1971. This led to the IEEE 1139 Standards that have served the community well, with advances and revisions continuing since their initial publication. Rubiola's book, based on his extensive seminar notes, is a capstone tutorial on the theoretical basis and experimental measurements of oscillators for which phase noise and frequency stability are primary issues.

In his first chapter Rubiola introduces the reader to the fundamental statistical descriptions of oscillator instabilities and discusses their role in the standards. Then in the second chapter he provides an exposition of the sources of noise in devices and circuits. In an instructive analysis of cascaded stages, he shows that, for modulative or parametric flicker noise, the effect of cascaded stages is cumulative without regard to stage gain.

This is in contrast with the well-known treatment of additive noise using the Friis formula to calculate an equivalent input noise power representing noise that may originate anywhere in a cascade of real amplifiers. This example highlights the concept that "the model is not the actual thing." He also describes concepts for the reduction of flicker noise in amplifier stages.

In his third chapter Rubiola then combines the elements of the first two chapters to derive models and techniques useful in characterizing phase noise arising in resonator feedback oscillators, adding mathematical formalism to these in the fourth chapter. In the fifth chapter he extends the reader's view to the case of delay-line oscillators such as lasers. In his sixth chapter, Rubiola offers guidance for the instructive "hacking" of existing oscillators, using their external phase spectra and other measurables to estimate their internal configuration. He details cases in which resonator fluctuations mask circuit noise, showing that separately quantifying resonator noise can be fruitful and that device noise figure and resonator Q are not merely arbitrary fitting factors.

It's interesting to consider what lies ahead in this field. The successes of today's consumer wireless products, cellular telephony, WiFi, satellite TV, and GPS, arise directly from the economies of scale of highly integrated circuits. But at the same time this introduces compromises for active-device noise and resonator quality. A measure of the market penetration of multi-signal consumer systems such as cellular telephony and WiFi is that they attract enough users to become interference-limited, often from subscribers much nearer than a distant base station. Hence low phase noise remains essential to preclude an unacceptable decrease of dynamic range, but it must now be achieved within narrower bounds on the available circuit elements.

A search for new understanding and techniques has been spurred by this requirement for low phase noise in oscillators and synthesizers whose primary character is integration and its accompanying minimal cost. This body of knowledge is advancing through a speculative and developmental phase. Today, numerical nonlinear circuit analysis supports additional design variables, such as the timing of the current pulse in nonlinear oscillators, that have become feasible because of the improved capabilities of both semiconductor devices and computers.

The field is alive and well, with emerging players eager to find a role on the stage for their own scenarios. Professionals and students, whether senior or new to the field so ably described by Rubiola, will benefit from his theoretical rigor, experimental viewpoint, and presentation.

David B. Leeson
Stanford University
February 2008

Preface

The importance of oscillators in science and technology can be outlined by two milestones. The *pendulum*, discovered by Galileo Galilei in the sixteenth century, persisted as "the" time-measurement instrument (in conjunction with the Earth's rotation period) until the piezoelectric quartz resonator. Then, it was not by chance that the first *integrated circuit*, built in September 1958 by Jack Kilby at the Bell Laboratories, was a radio-frequency oscillator.

Time, and equivalently frequency, is the most precisely measured physical quantity. The wrist watch, for example, is probably the only cheap artifact whose accuracy exceeds 10^{-5}, while in primary laboratories frequency attains the incredible accuracy of a few parts in 10^{-15}. It is therefore inevitable that virtually all domains of engineering and physics rely on time-and-frequency metrology and thus need reference oscillators. Oscillators are of major importance in a number of applications such as wireless communications, high-speed digital electronics, radars, and space research. An oscillator's random fluctuations, referred to as noise, can be decomposed into amplitude noise and phase noise. The latter, far more important, is related to the precision and accuracy of time-and-frequency measurements, and is of course a limiting factor in applications.

The main fact underlying this book is that an oscillator turns the phase noise of its internal parts into frequency noise. This is a necessary consequence of the Barkhausen condition for stationary oscillation, which states that the loop gain of a feedback oscillator must be unity, with zero phase. It follows that the phase noise, which is the integral of the frequency noise, diverges in the long run. This phenomenon is often referred to as the "Leeson model" after a short article published in 1966 by David B. Leeson [63]. On my part, I prefer the term *Leeson effect* in order to emphasize that the phenomenon is far more general than a simple model. In 2001, in Seattle, Leeson received the W. G. Cady award of the IEEE International Frequency Control Symposium "for clear physical insight and [a] model of the effects of noise on oscillators."

In spring 2004 I had the opportunity to give some informal seminars on noise in oscillators at the NASA/Caltech Jet Propulsion Laboratory. Since then I have given lectures and seminars on noise in industrial contexts, at IEEE symposia, and in universities and government laboratories. The purpose of most of these seminars was to provide a *tutorial*, as opposed to a report on advanced science, addressed to a large-variance audience that included technicians, engineers, Ph.D. students, and senior scientists. Of course, capturing the attention of such a varied audience was a challenging task. The stimulating discussions that followed the seminars convinced me I should write a working document[1] as a preliminary step and then this book. In writing, I have made a serious effort to address the same broad audience.

This work could not have been written without the help of many people. The gratitude I owe to my colleagues and friends who contributed to the rise of the ideas contained in this book is disproportionate to its small size: Rémi Brendel, Giorgio Brida, G. John Dick, Michele Elia, Patrice Féron, Serge Galliou, Vincent Giordano, Charles A. (Chuck) Greenhall, Jacques Groslambert, John L. Hall, Vladimir S. (Vlad) Ilchenko, Laurent Larger, Lutfallah (Lute) Maleki, Andrey B. Matsko, Mark Oxborrow, Stefania Römisch, Anatoliy B. Savchenkov, François Vernotte, Nan Yu.

Among them, I owe special thanks to the following: Lute Maleki for giving me the opportunity of spending four long periods at the NASA/Caltech Jet Propulsion Laboratory, where I worked on noise in photonic oscillators, and for numerous discussions and suggestions; G. John Dick, for giving invaluable ideas and suggestions during numerous and stimulating discussions; Rémi Brendel, Mark Oxborrow, and Stefania Römisch for their personal efforts in reviewing large parts of the manuscript in meticulous detail and for a wealth of suggestions and criticism; Vincent Giordano for supporting my efforts for more than 10 years and for frequent and stimulating discussions.

I wish to thank some manufacturers and their local representatives for kindness and prompt help: Jean-Pierre Aubry from Oscilloquartz; Vincent Candelier from RAKON (formerly CMAC); Art Faverio and Charif Nasrallah from Miteq; Jesse H. Searles from Poseidon Scientific Instruments; and Mark Henderson from Oewaves.

Thanks to my friend Roberto Bergonzo, for the superb picture on the front cover, entitled "The amethyst stairway." For more information about this artist, visit the website http://robertobergonzo.com.

Finally, I wish to thank Julie Lancashire and Sabine Koch, of the Cambridge editorial staff, for their kindness and patience during the long process of writing this book.

How to use this book

Let us first abstract this book in one paragraph. Chapter 1 introduces the language of phase noise and frequency stability. Chapter 2 analyzes phase noise in amplifiers, including flicker and other non-white phenomena. Chapter 3 explains heuristically the physical mechanism of an oscillator and of its noise. Chapter 4 focuses on the mathematics that describe an oscillator and its phase noise. For phase noise, the oscillator turns out to be a linear system. These concepts are extended in Chapter 5 to the delay-line oscillator and to the laser, which is a special case of the latter. Finally, Chapter 6 analyzes in depth a number of oscillators, both laboratory prototypes and commercial products. The analysis of an oscillator's phase noise discloses relevant details about the oscillator.

There are other books about oscillators, though not numerous. They can be divided into three categories: books on radio-frequency and microwave oscillators, which generally focus on the electronics; books about lasers, which privilege atomic physics and classical

---

[1] E. Rubiola, *The Leeson Effect – Phase Noise in Quasilinear Oscillators*, February 2005, arXiv:physics/0502143, now superseded by the present text.

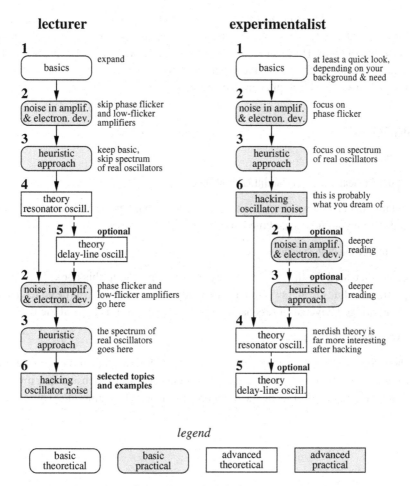

lecturer   experimentalist

legend

Figure 1 Asymptotic reading paths: on the left, for someone planning lectures on oscillator noise; on the right, for someone currently involved in practical work on oscillators.

optics; books focusing on the relevant mathematical physics. The present text is unique in that we look at the oscillator as a system consisting of more or less complex interacting blocks. Most topics are innovative, and the overlap with other books about oscillators or time-and-frequency metrology is surprisingly small. This may require an additional effort on the part of readers already familiar with the subject area.

The core of this book rises from my experimentalist soul, which later became convinced of the importance of the mathematics. The material was originally thought and drafted in the following (dis)order (see Fig. 1): **3** Heuristic approach, **6** Oscillator hacking, **4** Feedback theory, **5** Delay-line oscillators. The final order of subjects aims at a more understandable presentation. In seminars, I have often presented the material in the 3–6–4–5 order. Yet, the best reading path depends on the reader. Two paths are suggested in Fig. 1 for two "asymptotic" reader types, i.e. a lecturer and experimentalist. When planning to use this book as a supplementary text for a university course, the lecturer should be aware that students often lack the experience to understand and to appreciate Chapter 6 (Oscillator hacking) and other practical issues, while the theory can be more accessible to them. However, some mathematical derivations in Chapters 4 and 5 may require patience on the part of the experimentalist. The sections marked with one or two stars, ⋆ and ⋆⋆, can be skipped at first reading.

Supplementary material

My web page

http://rubiola.org   (also http://rubiola.net)

contains material covering various topics about phase noise and amplitude noise. A section of my home page, at the URL

http://rubiola.org/oscillator-noise

has been created for the supplementary material specific to this book. Oscillator noise spectra and slides from my seminars are ready. Other material will be added later.

Cambridge University Press has set up a web page for this book at the URL

www.cambridge.org/rubiola ,

where there is room for supplementary material. It is my intention to make the same material available on my home page and on the Cambridge website. Yet, my web page is under my full control while the other one is managed by Cambridge University Press.

Notation

The following notation list is not exhaustive. Some symbols are not listed because they are introduced in the main text. On occasion a listed symbol may have a different meaning, where there is no risk of ambiguity because the symbol has local scope and the usage is consistent with the general literature.

Uppercase is often used for

- Fourier or Laplace transforms
- constants, when the lower-case symbol is a function of time. For example, in relation to $v(t)$ we have V_{rms}, V_0 (peak)
- quantities conventionally represented with an upper-case symbol.
- in boldface, phasors. Example, $\mathbf{V} = V_{rms}e^{j\theta}$.

Though ω is the *angular frequency*, for short it is referred to as the frequency. Numerical values are always given in Hz. The symbol ω may be used as a shorthand for $2\pi\nu$ or $2\pi f$. The symbols ν and f always refer to single-sided spectra and ω always refers to two-sided spectra even if only the positive frequencies appear in plots.

Section 1.2 provides additional information about the relevant physical quantities, their meaning, and their usage, and about the variables associated with them.

The list includes some chapter, section, subsection, equation, or figure cross references.

Symbol	Meaning and text references
A	amplifier voltage gain (thus, the power gain is A^2)
b_i	coefficients of the power-law approximation of $S_\varphi(f)$. 1.6.2, (1.70), and Fig. 1.8
$b(t)$	resonator phase response. 4.4 and (4.62)
$\mathfrak{b}(t)$	resonator impulse response. 4.7
$B(s)$	resonator phase response, $B(s) = \mathcal{L}\{b(t)\}$. 4.4.2, 4.4.3
C	electrical capacitance, farad
$\mathcal{D}, \mathcal{D}(s)$	denominator (of a fraction or of a rational function)
E	energy, either physical (J) or mathematical (dimensionless), depending on context
$\mathcal{E}$	electric field, V/m
$\mathbb{E}$	mathematical expectation. 1.3.1 and (1.28)
f	Fourier frequency, Hz. 1.2
$f, f(x)$	generic function. 1.2

f_c	amplifier corner frequency, Hz. 2.3.3
f_L	Leeson frequency, Hz. 3.2 and (3.21)
F	amplifier noise figure. 2.2 and (2.11)
$\mathcal{F}\{\cdot\}$	Fourier transform operator. (A. 3)
h	Planck's constant, $h = 6.626 \times 10^{-34}$ J s
h_i	coefficients of the power-law approximation of $S_y(f)$. 1.6.3 and (1.73), (1.74)
$h(t)$	impulse response. 1.5.1 and (1.55), (1.56)
$\mathrm{h}(t)$	phase response
$H(s)$	transfer function, $H(s) = \mathcal{L}\{h(t)\}$, also $H(j\omega)$. 1.5.1 and (4.42)
$\mathrm{H}(s)$	phase transfer function, $\mathrm{H}(s) = \mathcal{L}\{\mathrm{h}(t)\}$, also $\mathrm{H}(j\omega)$ 3.2 and 4.5
$i(t)$	current, as a function of time
j	imaginary unit, $j^2 = -1$
k	Boltzmann constant, 1.381×10^{-23} J/K
$k_{\text{(subscript)}}$	a constant, k_d, k_o, k_L, etc.
l	harmonic order (in Chapter 5)
ℓ	voltage attenuation or loss (thus, the power loss is ℓ^2)
L	electrical inductance, H
$\mathcal{L}\{\cdot\}$	Laplace transform operator. 1.5.1 and (A.1)
$\mathscr{L}(f)$	single-sideband noise spectrum, dBc/Hz. 1.6.1 and (1.68)
m	integer (in Chapter 5)
m	modulation index (of light intensity)
$n(t)$	random noise, either near-dc or rf–microwave
N	integer
N	noise power spectral density, W/Hz
$\mathcal{N}, \mathcal{N}(s)$	numerator (of a fraction or of a rational function)
p	complex variable, replaces s when needed
$P, P(t)$	power, either physical (W) or mathematical (dimensionless), depending on context
q	electron charge, $q = 1.602 \times 10^{-19}$ C
Q	resonator quality factor. 4.1
R, R_0	resistance, load resistance (often $R_0 = 50\ \Omega$)
R	reflection coefficient. Chapter 5
$R(\tau)$	autocorrelation or correlation function. 1.4.1
s	complex variable, $s = \sigma + j\omega$
$S(f)$	power spectral density (PSD). 1.4.1, 1.4.2
$S_a(f)$	one-sided PSD of the quantity a
$S_\varphi(f)$	one-sided PSD of the random phase $\varphi(t)$. 1.6.1
$S^I(f)$	one-sided PSD. 1.4.2. The variable is could also be v
$S^{II}(\omega)$	two-sided PSD. (1.4.1).
t	time

T	equivalent noise temperature of a device. 2.2
T	observation or measurement time in truncated signals. 1.4.1
T	period, $T = 1/v$
T, T_0	absolute temperature, reference temperature $T_0 = 290$ K
T	transmission coefficient. 5.2
$U(t)$	Heaviside (step) function, $U(t) = \int \delta(t')\,dt'$
$v(t)$	voltage (in theoretical contexts, also a dimensionless signal)
$x, x(t)$	a generic variable
$x(t)$	phase-time fluctuation. 1.2 and (1.17)
$y(t)$	fractional-frequency fluctuation. 1.2 and (1.18)
V, V_0	dc or peak voltage
$V(s)$	Laplace transform of $v(t)$
$\mathbf{V}$	voltage phasor. 1.1
$\alpha(t)$	normalized amplitude noise. 1.1.1
$\beta(s)$	transfer function of the feedback path. 4.2 and Fig. 4.6
$\delta(t)$	Dirac delta function
Δ	difference operator, in Δv, $(\Delta\omega)(t)$, etc.
η	photodetector quantum efficiency. 2.2.3
θ	phase or argument of a complex function $\rho e^{j\theta}$
κ	small phase step. Chapter 4
λ	wavelength
μ	harmonic order in phase space. Chapter 5
v	frequency (Hz), used for carriers. 1.2
ρ	modulus of a complex function $\rho e^{j\theta}$
ρ	photodetector responsivity, A/W. 2.2.3
σ	real part of the complex variable $s = \sigma + j\omega$
$\sigma_y(\tau)$	Allan deviation, square root of the Allan variance $\sigma_y^2(\tau)$. 1.7
τ	measurement time, in $\sigma_y(\tau)$
τ	resonator relaxation time. 4.1
$\tau_{\mathrm{d}}, \tau_{\mathrm{f}}$	delay of a delay line, and group delay of the mode selector filter. Chapter 5
$\varphi, \varphi(t)$	phase (constant), phase noise. 1.1
$\Phi(s)$	phase noise, $\Phi(s) = \mathcal{L}\{\varphi(t)\}$
χ	dissonance. 4.2 and (4.31)
$\psi, \psi(t)$	amplifier static phase, phase noise. 3.2 and 4.5
$\Psi(s)$	amplifier phase noise, $\Psi(s) = \mathcal{L}\{\psi(t)\}$
ω	angular frequency, carrier or Fourier. 1.1
ω_0	oscillator angular frequency. 1.1
ω_{L}	Leeson angular frequency
ω_{n}	resonator natural angular frequency. 1.2
ω_{p}	resonator free-decay angular pseudo-frequency. 1.2
Ω	replaces ω, when needed
Ω	detuning angular frequency. Chapter 5

Subscript	Meaning
0	oscillator carrier, in ω_0, P_0, V_0, etc.
i	input. Examples $v_i(t)$, $\varphi_i(t)$, $\Phi_i(s)$
i	current. Example, shot noise $S_i(\omega) = 2q\bar{i}$
l	light
L	Leeson
L	loop
m	main branch
n	resonator natural frequency (ω_n, ν_n)
o	output. Examples $v_o(t)$, $\varphi_o(t)$, $\Phi_o(s)$
p	resonator free-decay pseudofrequency (ω_p, ν_p)
p	pole, as in $s_p = \sigma_p + j\omega_p$ (referring to a complex variable)
p	peak. Example, $V_p = \sqrt{2}V_{rms}$
rms	root mean square
z	zero, as in $s_z = \sigma_z + j\omega_z$ (referring to a complex variable)

Symbol	Meaning
$< >$	mean
$< >_N$	mean of N values. 1.3.1
$\overline{x}$	time average of x, for example. 1.3.1
$\leftrightarrow$	transform–inverse-transform pair. Example, $x(t) \leftrightarrow X(\omega)$
$*$	convolution. Example, $v_o(t) = h(t) * v_i(t)$. 1.5.1
$\asymp$	asymptotically equal

1 Phase noise and frequency stability

In *theoretical physics*, the word "oscillator" refers to a physical object or quantity oscillating sinusoidally – or at least periodically – for a long time, ideally forever, without losing its initial energy. An example of an oscillator is the classical atom, where the electrons rotate steadily around the nucleus. Conversely, in *experimental science* the word "oscillator" stands for an artifact that delivers a periodic signal, powered by a suitable source of energy. In this book we will always be referring to the artifact. Examples are the hydrogen maser, the magnetron of a microwave oven, and the swing wheel of a luxury wrist watch. Strictly, a "clock" consists of an oscillator followed by a gearbox that counts the number of cycles and the fraction thereof. In digital electronics, the oscillator that sets the timing of a system is also referred to as the clock. Sometimes the term "atomic clock" is improperly used to mean an oscillator stabilized to an atomic transition, because this type of oscillator is most often used for timekeeping.

A large part of this book is about the "precision"[1] of the oscillator frequency and about the mechanisms of frequency and phase fluctuations. Before tackling the main subject, we have to go through the technical language behind the word "precision," and present some elementary mathematical tools used to describe the frequency and phase fluctuations.

1.1 Narrow-band signals

The ideal oscillator delivers a signal

$$v(t) = V_0 \cos(\omega_0 t + \varphi) \qquad \text{(pure sinusoid)}, \tag{1.1}$$

where $V_0 = \sqrt{2}\, V_{\text{rms}}$ is the peak amplitude, $\omega_0 = 2\pi \nu_0$ is the angular frequency,[2] and φ is a constant that we can set to zero. Let us start by reviewing some useful representations associated with (1.1).

A popular way of representing a noise-free sinusoid $v(t)$ in Cartesian coordinates is the *phasor*, also called the *Fresnel vector*. The phasor is the complex number $\mathbf{V} = A + jB$ associated with $v(t)$ after factoring out the $\omega_0 t$ oscillation. The absolute value

---

[1] Here, the word "precision" is not yet used as a technical term.
[2] The symbol ω is used for the angular frequency. Whenever there is no ambiguity, we will omit the adjective "angular" and give the numerical value in Hz, which of course refers to $\nu_0 = \omega/(2\pi)$.

$|\mathbf{V}|$ is equal to the rms value of $v(t)$, and the phase arg $\mathbf{V}$ is equal to φ. The phase reference is set by $\cos \omega_0 t$. Alternatively, the phasor is obtained by expanding $v(t)$ as $V_0 (\cos \omega_0 t \cos \varphi - \sin \omega_0 t \sin \varphi)$. Then, the real part is identified with the rms value of the $\cos \omega_0 t$ component and the imaginary part with the rms value of the $- \sin \omega_0 t$ component. Thus, the ideal signal (1.1) may be represented as the phasor

$$\mathbf{V} = \frac{V_0}{\sqrt{2}} e^{j\varphi}$$

$$\text{(phasor)} . \tag{1.2}$$

$$\mathbf{V} = \frac{V_0}{\sqrt{2}} (\cos \varphi + j \sin \varphi)$$

A more powerful tool is the *analytic signal* $z(t)$ associated with $v(t)$, also called the *pre-envelope* and formally defined as

$$z(t) = v(t) + j\hat{v}(t) \qquad \text{(analytic signal)} , \tag{1.3}$$

where $\hat{v}(t)$ is the Hilbert transform of $v(t)$, i.e. $v(t)$ shifted by $90°$. The analytic signal is most often used to represent narrowband signals, i.e. signals whose power is clustered in a narrow band centered at the frequency ω_0. However, it is not formally required that the bandwidth be narrow, nor that the power be centered at ω_0, and not even that ω_0 be contained in the power bandwidth. Amplitude and phase can be (slowly) time-varying signals.

The analytic signal $z(t)$ is obtained from $v(t)$ by deleting the negative-frequency side of the spectrum and multiplying the positive-frequency side by a factor 2. Alternatively, $z(t)$ can be obtained using any of the following replacements:

$$v(t) = \frac{V_0}{\sqrt{2}} \cos(\omega_0 t + \varphi) \quad \Rightarrow \quad \begin{cases} z(t) = V(t) e^{j[\omega_0 t + \varphi(t)]} \\ z(t) = V(t) e^{j\varphi(t)} e^{j\omega_0 t} \\ z(t) = V(t)(\cos \varphi + j \sin \varphi) e^{j\omega_0 t} . \end{cases} \tag{1.4}$$

The analytic signal has two relevant properties.

1. A phase shift θ applied to $v(t)$ is represented as $z(t)$ multiplied by $e^{j\theta}$.
2. Since the power associated with negative frequencies is zero, the total signal power can be calculated using the positive frequencies only.

The *complex envelope* of $z(t)$, also referred to as the *low-pass* process associated with $z(t)$, is obtained by deleting the complex oscillation $e^{j\omega_0 t}$ in the analytic signal. The complex envelope is the natural extension of the phasor and is used when the amplitude and phase are allowed to vary with time:

$$\mathbf{V} = \frac{V_0}{\sqrt{2}} e^{j\varphi} \qquad \Longleftrightarrow \qquad \tilde{v}(t) = V(t) e^{j\varphi(t)} . \tag{1.5}$$

Strictly speaking, the phasor refers to a pure sinusoid. Yet the terms "phasor" and "time-varying phasor" are sometimes used in lieu of the term "complex envelope."

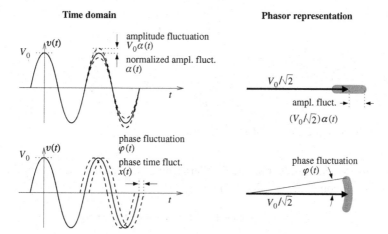

Figure 1.1 Amplitude and phase noise: V_0 is in volts, $\alpha(t)$ is non-dimensional, $\varphi(t)$ is in radians, and $x(t)$ is in seconds.

1.1.1   The clock signal

In the real world, an oscillator signal fluctuates in amplitude and phase. We introduce the quasi-perfect sinusoidal *clock signal* (Fig. 1.1)

$$v(t) = V_0[1 + \alpha(t)]\cos[\omega_0 t + \varphi(t)], \qquad |\alpha(t)| \ll 1, \ |\varphi(t)| \ll 1. \tag{1.6}$$

The term "clock signal" emphasizes the fact that the cycles of $v(t)$, and fractions thereof, can be counted by suitable circuits, so that $v(t)$ sets a time scale. When talking about clocks, we assume that (1.6) has a high signal-to-noise ratio. Hence we note the following.

- The peak amplitude V_0 of (1.1) is replaced by the envelope $V_0[1 + \alpha(t)]$, where $\alpha(t)$ is the random fractional amplitude.[3] The assumption

$$|\alpha(t)| \ll 1 \tag{1.7}$$

reflects the fact that actual oscillators have small amplitude fluctuations. Values $|\alpha(t)| \in (10^{-3}, 10^{-6})$ are common in electronic oscillators.
- The constant phase φ of (1.1) is replaced by the random phase $\varphi(t)$, which originates the clock error. In most cases, we can assume that

$$|\varphi(t)| \ll 1. \tag{1.8}$$

A slowly varying phase is often referred to as *drift*. Observing a clock in the long term, the assumption $|\varphi(t)| \ll 1$ is no longer true. Yet it is possible to divide the carrier frequency by a suitably large rational number. The phase scales down accordingly, so that the condition $|\varphi(t)| \ll 1$ is obtained at low frequencies.

---

[3] The symbol $\epsilon(t)$ is often used in the literature instead of $\alpha(t)$.

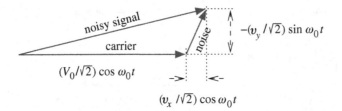

Figure 1.2 Phasor representation of a noisy sinusoid.

The clock signal can be rewritten in Cartesian coordinates by separating the $\cos \omega_0 t$ and $-\sin \omega_0 t$ components[4] (Fig. 1.2):

$$v(t) = V_0 \cos \omega_0 t + v_x(t) \cos \omega_0 t - v_y(t) \sin \omega_0 t . \qquad (1.9)$$

The signals $v_x(t)$ and $v_y(t)$ are called the *in-phase* and *quadrature* components of the noise, respectively.

The polar representation (1.6) and the Cartesian representation (1.9) are connected by

$$\alpha(t) = \sqrt{\left[1 + \frac{v_x(t)}{V_0}\right]^2 + \left[\frac{v_y(t)}{V_0}\right]^2} - 1 \qquad \text{(fractional amplitude)}, \qquad (1.10)$$

$$\varphi(t) = \arctan \frac{v_y(t)}{V_0 + v_x(t)} \qquad \text{(phase)} . \qquad (1.11)$$

Equation (1.10) is the Pythagorean theorem written in terms of the real component $(V_0 + v_x)/\sqrt{2}$ and the imaginary component $v_y/\sqrt{2}$. Equation (1.11) is arctan $\Im/\Re$, i.e. the arctangent of the imaginary-to-real ratio. A problem with (1.11) is that the arctangent returns the principal value, i.e. the value defined in $(-\pi/2, \pi/2)$, or in $(-\pi, \pi)$ using a two-argument arctangent. The cycles accumulated, if any, are to be counted separately. In low-noise conditions, it holds that, for $|v_x/V_0| \ll 1$ and $|v_y/V_0| \ll 1$,

$$\alpha(t) = \frac{v_x(t)}{V_0} \qquad \text{(fractional amplitude)} \qquad (1.12)$$

$$\varphi(t) = \frac{v_y(t)}{V_0} \qquad \text{(phase)} . \qquad (1.13)$$

The spectrum of a pure sinusoid such as (1.1) is an ideally thin line at ω_0, mathematically described as a Dirac delta function $\delta(\omega - \omega_0)$. Noise broadens the spectrum: the clock signal (1.6) looks like a line of bandwidth twice that of $\alpha(t)$ and $\varphi(t)$ if the signal-to-noise ratio is high. In the case of a low signal-to-noise ratio, $\varphi(t)$ yields a linewidth larger than twice the bandwidth.

Finally, the random phase $\varphi(t)$ does not contribute to the signal power. The instantaneous power is $P(t) = v^2(t)/R_0$. In low-noise conditions the power, averaged over

---

[4] The form $x(t) \cos \omega_0 t - y(t) \sin \omega_0 t$ is preferable for a signal in Cartesian coordinates, but in this chapter x and y are used for other relevant quantities.

a time T_m longer than the oscillation period yet shorter than the time scale T_α of the amplitude fluctuations, is

$$\overline{P(t)} = \frac{V_0^2}{2R_0}\left[1 + 2\alpha(t)\right], \qquad |\alpha(t)| \ll 1, \quad 2\pi/\omega_0 \ll T_m \ll T_\alpha . \qquad (1.14)$$

Example 1.1. Let us estimate the error accumulated in 1 year by a clock based on a 10 MHz oscillator accurate to within 10^{-10}. The maximum clock error is $T_e = (\Delta\omega/\omega)T_{meas}$; thus $T_e = 10^{-10} \times 3.16 \times 10^7$ s $= 3.2$ ms in one year. The oscillation period is $T_c = 2\pi/\omega_0 = 10^{-7}$ s. Hence the clock error accumulated in one year is $n = T_e/T_c = 3.16 \times 10^4$ cycles of the 10 MHz carrier; thus $\varphi = 2\pi n = 2 \times 10^5$ rad.

1.2   Physical quantities of interest

Traditionally, physicists use the symbol ν for the frequency while electrical engineers prefer f. It is unfortunate that in the domain of time-and-frequency metrology the notation is sometimes unclear or difficult to understand because both ν and f are found in the same context. When I came to metrology in the early 1980s with a background in electronics and telecommunications, it took me a long time to get used to this unnecessary complication. The early articles about frequency stability use ν for fixed frequencies, such as a carrier or a beat note, and f for the Fourier frequency, i.e. the variable of spectral analysis. Other articles use ν in the carrier signal $\cos 2\pi \nu_0 t$ and f for the spectral analysis of the low-pass fluctuations (α, φ, etc.), considering the carrier and the low-pass fluctuations as nearly separate worlds. Of course, these two distinctions between ν and f are similar, so it may be difficult to decide whether a frequency should be ν or f. Additional confusion arises from the fact that fluctuations even smaller than 10^{-16} are sometimes measured;[5] no frequency can be taken as constant. In the end, it is recommended that the exact meaning of a frequency symbol in a particular equation is always checked.

Another point is that the oscillator instability can be described as a phase fluctuation or as a frequency fluctuation. The first choice is made in the definition of the clock signal (1.6), here repeated:

$$v(t) = V_0[1 + \alpha(t)]\cos[\omega_0 t + \varphi(t)] .$$

In this representation, it is implicitly assumed that ω_0 is the best estimate of the oscillator frequency, so that $\omega_0 t$ describes the oscillation and $\varphi(t)$ describes its phase fluctuation. This approach is suitable for short-term measurements, where the oscillator stability is sufficient for $\varphi(t)$ to stay in the interval $(-\pi, \pi)$. In the longer term, the oscillator

---

[5] It is instructive to relate this small value to the internal computer representation of numbers. Since the IEEE standard "double precision" format has a 15 digit mantissa, the value 10^{-16} is one order of magnitude smaller than the roundoff error.

ends up drifting more than a half-cycle of the carrier frequency and the phase $\varphi(t)$ becomes ambiguous. In such cases, we may prefer to characterize the oscillation using the frequency fluctuations. Then the clock signal is written as

$$v(t) = V_0[1 + \alpha(t)]\cos\left[\omega_0 t + \int (\Delta\omega)(t)\, dt\right] \quad \text{(clock signal)}, \quad (1.15)$$

where

$$(\Delta\omega)(t) = \dot{\varphi}(t) \quad \text{(angular-frequency fluctuation)} \quad (1.16)$$

is the angular frequency fluctuation.

Additionally, it is often useful to normalize $\varphi(t)$ in order to transform the phase noise into time fluctuations, expressed in seconds (see below), and to normalize the oscillator frequency.

The definitions summarized below are aimed at giving straightforward access to the general literature on phase noise and frequency stability; Fig. 1.3 relates the quantities described to a typical experimental setup.

$\varphi(t)$ represents the phase noise, i.e. the random phase fluctuation defined by (1.6).

$\alpha(t)$ represents the fractional-amplitude noise (for short, "amplitude noise"), i.e. the random amplitude fluctuation defined by (1.6).

v is used for the carrier frequency, either radio, microwave, or optical, and also for the beat note between two carriers. The symbol v can be a variable, as on the frequency axis of a spectrum analyzer, or a constant, as in the nearly constant frequency of an oscillator. We can also find $v(t)$ used in the same way as the quantity $(\Delta v)(t)$ introduced below.

Example: let $v_1 = 10\,\text{GHz}$ and $v_2 = 10.24\,\text{GHz}$ be the frequency of two oscillators. On the spectrum analyzer, we see two lines at $v = v_1$ and $v = v_2$. After mixing, the beat frequency is $v_b = v_2 - v_1 = 240\,\text{MHz}$.

$\Delta v = v - v_0$ is the difference between the actual frequency v and a reference value v_0. The latter can be the nominal frequency or a reference value close to v.

$(\Delta v)(t)$ represents the instantaneous frequency fluctuation (or noise). This implies the assumption of slow modulation with a high modulation index, so that the signal can be approximated by a slow-swinging carrier. This means that the carrier and sidebands degenerate into a single Dirac δ function that tracks the modulation. In most practical cases, $(\Delta v)(t)$ should be regarded as the fluctuating output of a frequency comparator, after discarding the dc component.

f is used for frequency in the spectral analysis of low-pass processes, close to dc, after detection. Thus, f is used in connection with $\alpha(t)$, $\varphi(t)$, $v_x(t)$, $v_y(t)$, etc. The symbol f can refer to a variable, as on the frequency axis of a FFT[6] analyzer, or to a constant.

---

[6] There is no theoretical need to use a FFT (fast Fourier transform) analyzer to measure a near-dc process. However, this is the type of spectrum analyzer used in virtually all cases.

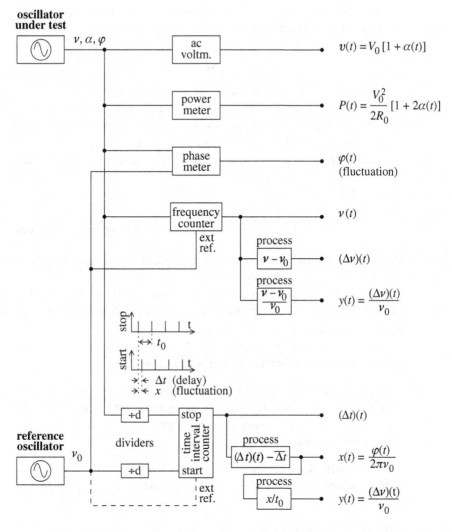

Figure 1.3 Simplified diagram of oscillator noise measurements in the time domain, illustrating the main physical quantities of interest in time-and-frequency metrology.

Example: inspecting the phase noise of a microwave amplifier in some specific conditions, we find white noise at high f and flicker noise below the corner frequency f_c.

ω is the angular frequency. The unit associated with ω is rad/s. In this book, ω is a shorthand for either $2\pi\nu$ or $2\pi f$. We also use $\Delta\omega$ and $(\Delta\omega)(t)$. The word "angular" is often omitted, and numerical values are given in Hz, which of course refers to $\omega/(2\pi)$.

Ω is used instead of ω in some special cases, as in Chapter 5, where we need to represent several angular frequencies at the same time. Preferentially, Ω refers to low-pass phenomena.

$x(t)$ is the phase-time fluctuation, that is, the random phase fluctuation $\varphi(t)$ converted into time, and measured in seconds:

$$x(t) = \frac{\varphi(t)}{\omega_0} = \frac{\varphi(t)}{2\pi \nu_0} \qquad \text{(phase-time fluctuation)} . \qquad (1.17)$$

Here ν_0 is either the nominal or the estimated frequency.

Interestingly, $x(t)$ does not become ambiguous when $\varphi(t)$ exceeds half a cycle of the carrier (Fig. 1.3) because it allows the accumulation of phase cycles.

$y(t)$ is the fractional-frequency fluctuation, i.e. the instantaneous frequency fluctuation normalized to the carrier frequency ν_0. The quantity $y(t)$ is dimensionless.

$$y(t) = \dot{x}(t) = \frac{\dot{\varphi}(t)}{\omega_0}$$

$$= \frac{(\Delta\omega)(t)}{\omega_0} = \frac{(\Delta\nu)(t)}{\nu_0} \qquad \text{(fractional-frequency fluctuation)} . \qquad (1.18)$$

The output of digital instruments is a stream of sampled values, denoted by an integer subscript. Thus, x_k is $x(t)$ sampled at the time $t = k\tau$.

Finally, a number of relationships are written in their usual form, found in most textbooks. Therefore, it is inevitable that some of the above symbols are also used in expressions such as "let $f(x)$ be a function . . . ," etc.

Three frequencies have a special rôle all through this book and are used extensively in Chapters 4 and 5; thus they deserve to be mentioned here.

ω_0 and ν_0 are the oscillator frequency and angular frequency, i.e. those of the carrier.

ω_n and ν_n are the natural angular frequency and natural frequency of a resonator.

ω_p and ν_p are the free-decay pseudofrequency and angular pseudofrequency of a resonator. It holds that $\omega_p \lesssim \omega_0$.

In most oscillators, the oscillation frequency ω_0 is determined by the natural frequency ω_n of a resonator. However, ω_0 differs slightly from ω_n because of feedback.

1.2.1   ⋆ Frequency synthesis

In a large number of applications the oscillator is the reference of a frequency synthesizer. A noise-free synthesizer can be regarded as a gearbox that multiplies the input frequency ω_i by a rational number $\mathcal{N}/\mathcal{D}$ and outputs a frequency

$$\omega_o = \frac{\mathcal{N}}{\mathcal{D}} \omega_i \qquad \left(\nu_o = \frac{\mathcal{N}}{\mathcal{D}} \nu_i \right) . \qquad (1.19)$$

In the presence of small fluctuations, the input fluctuations propagate to the synthesizer output with the same $\mathcal{N}/\mathcal{D}$ law, that is,

$$(\Delta\omega_o)(t) = \frac{\mathcal{N}}{\mathcal{D}} (\Delta\omega_i)(t) \qquad \left((\Delta\nu_o)(t) = \frac{\mathcal{N}}{\mathcal{D}} (\Delta\nu_i)(t) \right) , \qquad (1.20)$$

$$\varphi_o(t) = \frac{\mathcal{N}}{\mathcal{D}} \varphi_i(t) . \qquad (1.21)$$

A time lag can be present from input to output if the synthesizer includes a phase-locked loop (PLL). However, the fractional frequency fluctuation and the phase-time fluctuation observed at the output are equal to the input fluctuations:

$$y_o(t) = y_i(t),$$ (1.22)

$$x_o(t) = x_i(t).$$ (1.23)

Finally, we notice that there is no general law for the propagatation of the amplitude fluctuation $\alpha(t)$ through a synthesis chain. The reason is that the synthesis needs strong nonlinearity, hence the amplitude is saturated.

Large phase noise

The above rules hold when the phase noise is low. In large-phase-noise conditions, the synthesizer's behavior is governed by the energy conservation law. The easiest way to understand this is to write the output signal in the analytic form $z(t) = V_{rms}\, e^{j\omega_0 t}\, e^{j\varphi(t)}$. The phase-noise term $e^{j\varphi(t)}$ spreads the power into the noise sidebands, yet without changing the total power because $|e^{j\varphi(t)}| = 1$. If the output phase noise exceeds some 2 radians, the sidebands sink most of the power and the carrier power drops abruptly. This phenomenon is referred to as *carrier collapse*. As a consequence, extremely high spectral purity is needed when the multiplication ratio is high, for example in the synthesis of THz or optical signals from electronic oscillators.

Mathematically, carrier collapse is a consequence of the application of the Angers–Jacobi expansion

$$e^{jz\cos\alpha} = \sum_{n=-\infty}^{\infty} J_n(z)\, e^{jn\alpha}$$ (1.24)

to the angular modulation, according to which the carrier amplitude is dominated by the Bessel function $J_0(z)$. The function $J_0(z)$ nulls at $z \simeq 2.405$. Further consideration of this phenomenon is beyond our scope, however.

1.3 Elements of statistics

In the previous sections, we expressed the phase noise and amplitude noise in terms of simple time-dependent functions, denoted by $\alpha(t)$ and $\varphi(t)$. In reality, to address the nature of noise properly some statistical tools are necessary, and the concept of a random process needs to be introduced. A few definitions are given below to establish the vocabulary. The reader is encouraged to study the subject using appropriate references, among which I prefer [32, 74].

1.3.1   Basic definitions

Random or stochastic process
A *random process* is defined through a random experiment **e** that associates a time-domain function $x_e(t)$ with each outcome e. The specification of such an experiment, together with a probability law, defines a random process $\mathbf{x}(t)$. Each random process has an infinite number of *realizations*, which form an *ensemble*. A realization, also called a *sample function*, is a time-domain signal $x_e(t)$. For short, the subscript e is dropped whenever there is no ambiguity or no need to refer to a specific outcome e.

A random process and its associated ensemble are powerful mathematical concepts, but they are not directly accessible to the experimentalist, who can only measure a finite number of realizations.

Mean, time average, and expectation
In the measurement of random processes (subsection 1.3.2) we use simultaneously three types of "average," the simple mean, the time average, and the mathematical expectation. Hence, for clarity we need different notation for these.

Given a series of N data x_i, the simple *mean* of x is denoted by angle brackets:

$$\langle x \rangle_N = \frac{1}{N} \sum_{i=1}^{N} x_i \qquad \text{(mean)}. \tag{1.25}$$

The simple mean is often used to average the output stream of an instrument. The quantity x is unspecified. For example, we can average in this way a series of numbers, a series of spectra, etc.

The *time average* of x is denoted by an overbar:

$$\overline{x} = \frac{1}{T} \int_{-T/2}^{T/2} x(t)\,dt \qquad \text{(time average)} \tag{1.26}$$

In the case of causal systems, where the response starts at $t = 0$, the integration limits can be changed from $-T/2$ and $T/2$ to 0 and T. In most cases, the readout of an instrument is of the form (1.26). This means that the input quantity x is averaged uniformly over the time T.

More generally, the time average includes a weight function $w(t)$:

$$\overline{x} = \int_{-\infty}^{\infty} x(t)\,w(t)\,dt \qquad \text{(weighted time average)} \tag{1.27}$$

with

$$\int_{-\infty}^{\infty} w(t)\,dt = 1.$$

In theoretical discussions the definition (1.27) is generally adopted as the *definition of the measure of* x. The readout of sophisticated instruments can be of this type.

The statistical *expectation* $\mathbb{E}\{\mathbf{x}\}$ is the extension of the average to stochastic processes. It is defined as

$$\mathbb{E}\{\mathbf{x}\} = \int_{-\infty}^{\infty} x f_{\mathbf{x}}(x)\,dx \qquad \text{(expectation)}, \qquad (1.28)$$

where $f_{\mathbf{x}}(x)$ is the probability density of the process $\mathbf{x}(t)$. Of course, the expectation can be calculated for all the parameters of interest.

Stationarity and repeatability

A random process $\mathbf{x}(t)$ is said to be *stationary* (in a strict sense) if all its statistical properties are invariant under an arbitrary time shift, that is, $\mathbf{x}(t) \to \mathbf{x}(t + t')$. Of course, strict-sense stationarity can only be assessed a priori on the basis of physical laws. A less general invariance property is *wide-sense stationarity*. This definition requires that

1. the statistical expectation $\mathbb{E}\{\mathbf{x}(t)\}$ is constant,
2. the autocorrelation function $R(t', t'') = \mathbb{E}\{\mathbf{x}(t')\mathbf{x}(t'')\}$ depends only on the time difference $\tau = t' - t''$; thus, $R(\tau) = \mathbb{E}\{\mathbf{x}(t + \tau)\mathbf{x}(t)\}$.

Stationarity is closely related to the empirical concept of repeatability. In fact, repeating an experiment means doing it again with the same equipment and under the same conditions, yet inevitably at a different time. The *repeatability* is expressed by the interval containing the spread of measured values when the experiment is repeated a number of times.

Cyclostationary process

A random process $\mathbf{x}(t)$ is *cyclostationary* of period T if all its statistical properties are invariant under a time shift of nT, that is, $\mathbf{x}(t) \to \mathbf{x}(t + nT)$ for integer n. The property of cyclostationarity is an extension of the concept of periodicity and therefore is of paramount importance in radio engineering, where the information is carried by strong periodic signals. And of course it is of paramount importance in oscillators.

Ergodicity and reproducibility

The process $\mathbf{x}(t)$ is *ergodic* if all its statistical properties can be estimated from a single realization $x_e(t)$. In practice, ergodicity makes sense only for stationary or cyclostationary processes. It makes random processes accessible to the experimentalist. In fact, in the case of an ergodic process, the ensemble average is equal to the time average of a single realization of the process, which can be measured. The most important averages are summarized in Table 1.1.

The concept of *reproducibility* is closely connected with ergodicity. In physics, reproducing an experiment means that the same experiment is done by different operators in different locations but using a close copy of all the relevant parts. Of course, we expect that all the replicas of the experiment give the "same result." Using the vocabulary of statistics, each replica of the experiment is a realization (or sample) picked from the ensemble. The reproducibility is expressed by the spread of values of a replicated experiment.

Table 1.1 Relevant statistical and time-domain averages. With minor changes, the integration limits can be set to 0 and T

Parameter	Statistical expectation	Time-domain average
AVG (dc level)	$\mathbb{E}\{\mathbf{x}(t)\}$	$\overline{x(t)} =$ $$\lim_{T \to \infty} \frac{1}{T} \int_{-T/2}^{T/2} x(t)\,dt$$
second moment (power)	$\mathbb{E}\{|\mathbf{x}(t)|^2\}$	$\overline{|x(t)|^2} =$ $$\lim_{T \to \infty} \frac{1}{T} \int_{-T/2}^{T/2} |x(t)|^2\,dt$$
VAR (ac power)	$\mathbb{E}\{|\mathbf{x}(t) - \mathbb{E}\{\mathbf{x}(t)\}|^2\}$	$\overline{|x(t) - \overline{x(t)}|^2} =$ $$\lim_{T \to \infty} \frac{1}{T} \int_{-T/2}^{T/2} |x(t) - \overline{x(t)}|^2\,dt$$
auto-correlation function $R(\tau)$	$\mathbb{E}\{\mathbf{x}(t)\,\mathbf{x}(t + \tau)\}$	$\overline{x(t)\,x(t + \tau)} =$ $$\lim_{T \to \infty} \frac{1}{T} \int_{-T/2}^{T/2} x(t)\,x(t + \tau)\,dt$$

Example 1.2. Thermal noise of a resistor. Suppose that we measure the variance σ^2 of the thermal voltage $v(t)$ across a resistor R in a bandwidth B, while the resistor is in thermal equilibrium at a temperature T. We want to establish the properties of the random process associated with this experiment. In terms of statistics, the choice of a particular resistor and of a specific time interval (t_1, t_2) is the outcome e of the random experiment $\mathbf{e}$. The waveform $v_e(t)$ that we record for a duration from t_1 to t_2 is a realization. The measured variance σ_e^2 is the second moment of $v_e(t)$.

First, we restrict our attention to a single resistor and identify a sub-process, its thermal noise. We measure the latter a number of times for the same set of conditions. All the measurements, which differ only in the time interval, will give the result $\sigma^2 = 4kTRB$, the same within a confidence interval. So we can state that the sub-process (the thermal noise of the specific resistor) is stationary.

Second, we measure many resistors with the same resistance R in the same conditions and simultaneously. Suppose that in all cases we obtain the value $\sigma^2 = 4kTRB$, the same as that obtained using one resistor on successive time slots. This means that the time average is equal to the ensemble average and hence that the process is ergodic.

Third, we observe that the law $\sigma^2 = 4kTRB$ holds for all resistors at any time, thus it describes the entire process. Thus the process is stationary and ergodic.

Finally, we notice that the variance σ^2 decreases from $4kTRB$ beyond the cutoff frequency $v = kT/h$ at which the photon energy hv equals the thermal energy kT. An appropriate decrease (roll-off) is necessary for the total power to be finite.

1.3.2 The measurement of random processes

With reference to Table 1.1, most quantities measured in a laboratory are estimates of the expectation $\mathbb{E}\{p\}$ of a function p associated with a random process $\mathbf{x}(t)$. Such measured quantities could be the average, the variance, the autocorrelation function, etc. We rely on the assumption that $\mathbf{x}(t)$ is stationary and ergodic with regard to p. However complex the mathematical background might be, the experimental process turns out to be simple. We follow the steps detailed below.

1. We collect N measures $\overline{p}_i$, that is, N averages of the quantity p on time slots of duration T centered at t_i:

$$\overline{p}_i = \frac{1}{T} \int_{-T/2+t_i}^{T/2+t_i} p(t)\,dt, \qquad i \in [1, N]. \tag{1.29}$$

The time slots can be either far from one another, contiguous, or partially overlapped. For each random process, the number of degrees of freedom is related to N and to the dead time between the measurements.

2. We use the simple mean $\langle p \rangle_N$, which is an average over the realizations of the process, as an estimator of the quantity p:

$$\hat{p} = \frac{1}{N} \sum_{i=1}^{N} p_i, \tag{1.30}$$

3. We trust the mean as an unbiased estimator:

$$\hat{p} \to \mathbb{E}\{p\}. \tag{1.31}$$

The expected discrepancy decreases on increasing the number of degrees of freedom.

1.3.3 Noise in RF and microwave circuits

A physical experiment can be trusted only if it is repeatable and reproducible. Therefore, a lack of stationarity is often regarded as the result of a mistake. Non-stationary results are found, for example, if the experiment or its boundary conditions vary in time. Radio-frequency circuits are no exception to this rule. Yet, often the word "stationary" should be replaced by *cyclostationary*. This is typical of large-signal circuits, dominated by a strong carrier. In such cases some *hidden* stationary process modulates the carrier, thus the *visible* process is cyclostationary. In some more complex cases, the noise and carrier interact because the carrier modulates the noise parameters. This will be addressed in Chapter 2 (on phase noise in semiconductors and amplifiers).

1.4 The measurement of power spectra

The most familiar example of a power spectrum occurs when white light is turned into a continuous distribution of rainbow colors by a prism. This experiment can tell us how the light power is distributed on the frequency axis or on the wavelength axis. Another

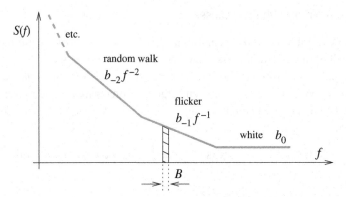

Figure 1.4 Power spectral density $S(f)$. The PSD can be regarded as the power $P = \lim_{B \to 0} B^{-1} \int_{f-B/2}^{f+B/2} S(f') \, df'$. The noise types (white, flicker, etc.) are those commonly encountered in the measurement of oscillator phase noise.

example familiar to the experimentalist is the spectrum analyzer. This instrument is a tunable filter that measures the power in a bandwidth B centered at a frequency f, which is swept.

As a *physical quantity*, the one-sided power spectral density[7] $S_x(f)$ associated with a random process $x(t)$ can be operationally defined as a distribution[8] having the following properties.

1. The integral over the interval (a, b) is the signal power in that interval:

$$P_{ab} = \int_a^b S(f) \, df \,. \tag{1.32}$$

This expression enables us to extract a quasi-monochromatic beam from white light and gives the basis of the monochromator. Moreover, we can analyze the light, describe the dominance of the various colors, and identify the signature of atoms, ions, and molecules. This is the foundation of spectroscopy.

2. The power in two separate frequency slots (a, b) and (c, d) adds up:

$$P_{ab} + P_{cd} = \int_a^b S(f) \, df + \int_c^d S(f) \, df \,. \tag{1.33}$$

3. The integral over all frequencies is the total power of the signal:

$$P = \int_0^\infty S(f) \, df \,. \tag{1.34}$$

This is necessary for energy conservation.

In a simplistic way, $S_x(f)$ can be regarded as the average power of $x(t)$ in a 1 Hz bandwidth centered at f, yet measured on an ideally narrow frequency slot in order to remove the effect of slope (Fig. 1.4).

---

[7] If there is no ambiguity in the identification of the process, the subscript can be dropped.
[8] In mathematics, the distribution is an extension of the concept of a function. A distribution can only be defined through an integral, which is our case.

1.4.1 Theoretical background

While the physical approach to the power spectrum is intuitive and straightforward, the formal definition requires a complex theoretical framework that combines generalized Fourier analysis and the theory of random processes. We provide only some relevant results, and advise the reader to refer to specialized books for the details. The author's favorite references are [9, 56, 75].

In the case of a deterministic signal, the spectrum is simply defined as the squared modulus of the Fourier transform. In the case of random signals, things are a little different. The power spectrum $S(\omega)$ of a random process $\mathbf{x}(t)$, here called the *two-sided* power spectrum and denoted by $S^{II}(\omega)$, is formally defined as the Fourier transform of the autocorrelation function $R(\tau)$:

$$S^{II}(\omega) = \int_{-\infty}^{\infty} R(t) e^{-j\omega t} \, dt \qquad \text{(definition of } S^{II}(\omega)) , \tag{1.35}$$

$$R(\tau) = \mathbb{E}\{\mathbf{x}(t)\mathbf{x}(t+\tau)\} \qquad \text{(definition of } R(\tau)) . \tag{1.36}$$

We restrict our attention to the case of a *stationary* and *ergodic* random process $\mathbf{x}(t)$, accessible through a realization $x(t)$. All averages can be calculated from the realization $x(t)$; thus

$$R(\tau) = \lim_{T \to \infty} \frac{1}{T} \int_{-T/2}^{T/2} x(t)x(t+\tau) \, dt . \tag{1.37}$$

The direct measurement of $R(\tau)$ is often difficult. A more convenient way to estimate $S(\omega)$ is based on

$$S^{II}(f) = \lim_{T \to \infty} \frac{1}{T} |X_T(f)|^2 , \tag{1.38}$$

which follows from the Wiener–Khintchine theorem for ergodic stationary processes. The function $X_T(\omega)$ is the Fourier transform of the truncated signal $x_T(t)$:

$$X_T(\omega) = \int_{-T/2}^{T/2} x(t)e^{-j\omega t} \, dt , \tag{1.39}$$

where the integral limits indicate a finite observation time for $x(t)$.

According to the rules given in subsection 1.3.2 above, the power spectrum is measured by averaging m realizations:

$$S^{II}(f) = \frac{1}{T} \left\langle |X_T(f)|^2 \right\rangle_m . \tag{1.40}$$

1.4.2 One-sided and two-sided spectra

The Fourier transform–inverse-transform pair

$$X(\omega) = \int_{-\infty}^{\infty} x(t)e^{-j\omega t} dt \qquad \leftrightarrow \qquad x(t) = \frac{1}{2\pi} \int_{-\infty}^{\infty} X(\omega)e^{j\omega t} d\omega \tag{1.41}$$

yields naturally a frequency representation on negative and positive frequencies. As the actual waveforms $x(t)$ are real signals, $X(\omega)$ has the following symmetry properties:

$$\Re\{X(\omega)\} = \Re\{X(-\omega)\} \qquad \text{(the real part is even)} \tag{1.42}$$

$$\Im\{X(\omega)\} = -\Im\{X(-\omega)\} \qquad \text{(the imaginary part is odd)} \tag{1.43}$$

$$|X(\omega)| = |X(-\omega)| \qquad \text{(the modulus is even)} \tag{1.44}$$

$$\arg X(\omega) = -\arg X(-\omega) \qquad \text{(the argument (phase) is odd)} \tag{1.45}$$

Thus, the two-sided power spectrum is an even function of ω,

$$S_x^{II}(\omega) = S_x^{II}(-\omega), \tag{1.46}$$

and, as a consequence, *the negative-frequency half-plane is redundant*. A spectrum analyzer displays the one-sided spectrum $S^I(\omega)$:

$$S^I(\omega) = \begin{cases} 2S^{II}(\omega) & f > 0, \\ 0 & f < 0, \end{cases} \tag{1.47}$$

where the factor 2 is introduced to preserve the total power. However, *virtually all the literature on theoretical concerns uses the two-sided representation*, with positive and negative frequencies, because the formalism is clearer.

The relationship between S^I and S^{II} is imposed by the equivalence of the power in the frequency interval (a, b):

$$P_{ab} = \frac{1}{2\pi} \int_a^b S^I(\omega) \, d\omega, \tag{1.48}$$

$$P_{ab} = \frac{1}{2\pi} \int_{-b}^{-a} S^{II}(\omega) \, d\omega + \frac{1}{2\pi} \int_a^b S^{II}(\omega) \, d\omega. \tag{1.49}$$

As a general rule, we prefer ω for theoretical discussion, and f or ν for experiments, and thus ω for two-sided spectra and f or ν for one-sided spectra. Accordingly, we use

$$\omega, S^{II}(\omega) \qquad \qquad \text{in theoretical contexts},$$
$$\nu, S^I(\nu), \quad f, S^I(f) \qquad \text{in experimental contexts},$$

dropping the superscripts I and II unless there is a risk of ambiguity.

1.4.3 Spectrum analyzers

Table 1.2 shows the physical quantities associated with the spectral representation of $x_T(t)$. The Fourier transform is preferred in theoretical contexts but the one-sided power spectral density $S^I(f) = (2/T)|X_T(f)|^2$ is most suitable when measuring the noise spectrum. The one-sided quantity $(2/T^2)|X_T(f)|^2$ is preferred for the power measurement of carriers, i.e. sinusoidal signals.

Table 1.2 Physical quantities associated with the spectral representation of a random voltage $x(t)$, where T is the measurement time

Quantity	Physical dimension	Purpose
$X_T(f)$	V/Hz	theoretical discussion
$S^I(f) = \dfrac{2}{T}|X_T(f)|^2, \quad f > 0$	V^2/Hz or W/Hz	measurement of noise level (power spectral density)
$\dfrac{1}{T}S^I(f) = \dfrac{2}{T^2}|X_T(f)|^2, \quad f > 0$	V^2 or W	power measurement of carriers (sinusoidal signals)

In some literature, as well as in the technical documentation of some instruments, one finds the following terminology:

$$S^I(f) = \frac{2}{T}|X_T(f)|^2 \qquad \text{(power spectral density (PSD), V}^2\text{/Hz)} \qquad (1.50)$$

$$\frac{1}{T}S^I(f) = \frac{2}{T^2}|X_T(f)|^2 \qquad \text{(power spectrum (PS), V}^2\text{)} . \qquad (1.51)$$

However, *in most general literature the terms spectrum, power spectrum, and power spectral density are considered synonymous and used interchangeably*. We advise that the experimentalist relies on the measurement unit (V^2/Hz or V^2), rather than on the terms.

There are two basic types of spectrum analyzer for electrical signals, the traditional RF or microwave spectrum analyzer and the FFT analyzer. As the technology progresses, more digital technology is included in the RF or microwave instruments and the difference becomes smaller.

RF or microwave spectrum analyzer

This type of instrument is a superheterodyne receiver, swept in frequency, followed by a power meter. The sweep signal must be slow enough for the intermediate-frequency (IF) filter to respond stationarily. The physical quantity measured by this instrument is the power P dissipated by the input resistance R_0 in the bandwidth B centered at the frequency v, and is expressed in watts. The bandwidth B is determined by the IF filter, which can be selected from the front panel. The following cases deserve attention.

> *Pure sinusoid of frequency v_0 and power P_0.* The analyzer displays a (narrow) lorentzian-like distribution of width B centered at the frequency v_0. The peak is at the power P_0, while the actual shape of the distribution gives the frequency response of the IF filter. Thus the displayed linewidth depends on the choice of B, while the displayed v_0 and P_0 values are those of the input signal.
>
> *White noise of power spectral density N.* The analyzer displays a power level $P = NB$, which depends on the choice of B.

In principle, it is always possible to convert the reading (the total power in the bandwidth B of the IF filter) into the power spectral density by dividing by B. In old fully analog instruments there can be a loss of accuracy due to the large uncertainty in B. Furthermore the filter bandwidth can change with frequency and temperature, and it can age with time. Conversely, in modern instruments B is defined numerically after digitizing the signal, for there is no accuracy degradation due to the change in unit.

FFT spectrum analyzers

In this type of analyzer, the spectrum is inferred from a digitized time series of the input signal using

$$S^I(f) = \frac{2}{T} |X_T(f)|^2, \qquad f > 0. \tag{1.52}$$

The Fourier transform is therefore replaced by a fast Fourier transform (FFT). The reader should refer to [14] for a detailed account of the discrete Fourier transform and the FFT algorithm.

Hard truncation of the input signal yields the highest frequency resolution for a given T and a given number of samples in the time series. The problem with hard truncation is that the equivalent filter has high secondary lobes, owing to the Fourier transform of the rectangular pulses. In consequence, a point on the frequency axis is polluted by the signal (or noise) present in other parts of the spectrum. This phenomenon is called "frequency leakage." The leakage is reduced by replacing hard truncation with a weight function that in this context is called a *window*. The most popular such functions are the Hanning (cosine) window, the Bartlett (triangular) window, and the Parzen window. Hard truncation of the observation time gives a flat-top window. The choice of the most appropriate window function is a trade-off between accuracy and frequency leakage that depends on the specific spectrum.

The FFT, inherently, is linear in amplitude and frequency even if the spectrum is displayed on a logarithmic scale. The bandwidth B is determined by the frequency span and by the number of points of the FFT algorithm that are implemented. A typical analyzer has 1024 points. The lower 801 points are displayed, while the upper 223 points are discarded internally because they have been corrupted by aliasing. Thus, for example, using a span of 100 kHz the bandwidth is $10^5/800 = 125$ Hz.

The measured spectrum can be displayed in two ways, listed below, without loss of accuracy in the conversion.

Power spectral density $S^I(f) = (2/T)|X_T(f)|^2$ (V²/Hz). This choice gives a straightforward representation of the noise. White noise of power spectral density N (V²/Hz) is shown as a horizontal line of value N. Conversely, a pure sinusoid of amplitude V_{rms} and frequency f_0 appears as a (narrow) lorentzian-like distribution of width B centered at the frequency f_0. The peak is equal to V_{rms}^2/B, while the actual shape of the distribution is determined by the instrument's internal algorithm.

Power spectrum $(1/T)S^I(f) = (2/T^2)|X_T(f)|^2$ (V²). The spectrum is displayed as in an RF spectrum analyzer but with the trivial difference that the power is given in V² instead of in W. The input resistance is assumed to be 1 Ω unless otherwise specified. Thus a sinusoidal signal of amplitude V_{rms} is shown as a distribution of width B and height V_{rms}^2, while white noise of power spectral density N (V^2/Hz) is shown as a horizontal line at $V_{rms}^2 = NB$.

1.5   Linear and time-invariant (LTI) systems

1.5.1   System function

The analysis of a physical system can be reduced to the analysis of its input–output *transfer function*, also called the *system function*, here denoted by the operator L: $v_o(t) = L\{v_i(t)\}$. The system is assumed to be (at least locally) linear, so that

$$L\{av_i\} = aL\{v_i\}$$
$$L\{v_1 + v_2\} = L\{v_1\} + L\{v_2\} \qquad \text{(linearity)}, \qquad (1.53)$$

and time invariant, i.e. governed by parameters that are constant in time, so that

$$v_o(t) = L\{v_i(t)\} \quad \Leftrightarrow \quad v_o(t+t') = L\{v_i(t+t')\} \qquad \text{(time invariance)}. \quad (1.54)$$

Such systems, called linear and time-invariant (LTI) systems, have the amazing property that they can be described completely by their response, denoted by $h(t)$ to the Dirac delta function $\delta(t)$. Therefore, the output from an arbitrary input is

$$v_o(t) = h(t) * v_i(t) \qquad \text{(convolution)} \qquad (1.55)$$

$$= \int_{-\infty}^{+\infty} v_i(\tau) h(t-\tau) \, d\tau , \qquad (1.56)$$

and in the frequency domain

$$V_o(j\omega) = H(j\omega) V_i(j\omega) ; \qquad (1.57)$$

note that an upper-case letter denotes the Fourier transform of the corresponding lower-case function of time. Figure 1.5 summarizes some facts about the transfer function.

Another relevant property of LTI systems is that the exponential $e^{j\omega t}$ is an eigenfunction, hence the system response to $v_i(t) = e^{j\omega t}$ is $v_o(t) = Ce^{j\omega t}$. This follows immediately from (1.57). It turns out that the complex constant C is equal to $H(j\omega)$. Hence

$$v_i(t) = e^{j\omega t} \quad \Leftrightarrow \quad v_o(t) = H(j\omega) e^{j\omega t} . \qquad (1.58)$$

The function $H(j\omega)$ is often plotted as $20 \log_{10} |H(j\omega)|$ (dB) and arg $H(j\omega)$ (degrees or radians) on a logarithmic frequency axis; this is referred to as a Bode plot.

When the input is a random process, we have to replace the Fourier transforms $V_i(j\omega)$ and $V_o(j\omega)$ by the power spectra $S_i(\omega)$ and $S_o(\omega)$. Thus, (1.57) turns into

$$S_o(\omega) = |H(j\omega)|^2 S_i(\omega) . \qquad (1.59)$$

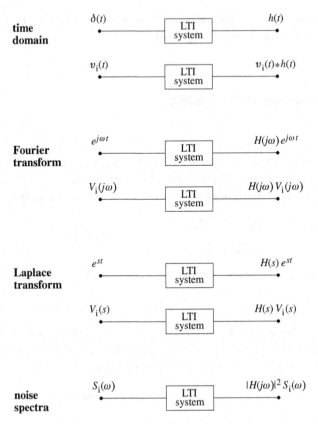

Figure 1.5 Some relevant properties of the transfer function in LTI systems.

It is worth mentioning that the system function $H(j\omega)$ is always a two-sided Fourier transform even if the one-sided spectra S_i and S_o are preferred. This is obvious from (1.59), because the use of single-sided spectra introduces a factor 2 into both $S_i(\omega)$ and $S_o(\omega)$, and thus $H(j\omega)$ must not be changed.

1.5.2   Laplace transform

The *Laplace transform* is generally preferred to the Fourier transform in circuit theory, and it is extensively used in Chapters 4 and 5. The system functions are taken to be causal, that is, $h(t) = 0$ for $t < 0$. This expresses the obvious fact that the output is zero before the input stimulus. Thus, denoting the system function by $x(t)$, the Fourier transform

$$X(j\omega) = \int_{-\infty}^{\infty} x(t)e^{-j\omega t}\, dt \qquad \text{(Fourier transform)} \qquad (1.60)$$

becomes

$$X(j\omega) = \int_0^\infty x(t)e^{-j\omega t}\, dt \qquad \text{for } x(t) = 0, \ \forall t < 0 \qquad \text{(causal signal)} . \quad (1.61)$$

Replacing $j\omega$ by $s = \sigma + j\omega$, we get the Laplace transform

$$X(s) = \int_0^\infty x(t)e^{-st}\, dt \qquad \text{(Laplace transform)} . \qquad (1.62)$$

Thus, the Laplace transform differs from the Fourier transform in that the frequency is allowed to be complex, while the strong-convergence problem arising from the lower integration limit $t = -\infty$ is eliminated by the assumption that $x(t) = 0$ for $t < 0$. A short summary of useful properties of and formulae for the Laplace transform is given in Appendix A.

The main reason for preferring the Laplace transform is that it gives access to the world of analytic functions. Analytic functions have the very important property of being completely determined by the positions of their roots (poles and zeros) on the complex plane and by the pole residues. This simplifies network analysis and synthesis dramatically.

1.5.3 Slow-varying systems

The system function $h(t)$, as any other physical quantity, can only be observed over a finite time, which we denote by T; of course, this observation time must be long enough to include (almost) all the energy of $h(t)$. Accordingly, the Fourier and Laplace transforms can only be evaluated over the time T:

$$H_T(j\omega) = \int_0^T h(t)e^{-j\omega t}\, dt , \qquad (1.63)$$

$$H_T(s) = \int_0^T h(t)e^{-st}\, dt . \qquad (1.64)$$

From this standpoint, a time-varying system can be seen as time invariant, provided that the time scale of the fluctuation is sufficiently slow compared with the observation time T. In the case of the Laplace transform, we can describe a slow-varying system in terms of *roots* that *fluctuate or drift on the complex plane*. This is an appropriate way to describe a fluctuating or drifting resonator.

Generalizing the finite observation time consists of introducing in the integral a weight function $w(t)$ of duration T, which sets the measurement time and has more or less smooth boundaries. Thus, the Laplace transform of the weighted signal is

$$H_w(j\omega) = \int_0^\infty h(t)w(t)e^{-j\omega t}\, dt , \qquad (1.65)$$

$$H_w(s) = \int_0^\infty h(t)w(t)e^{-st}\, dt . \qquad (1.66)$$

It may be remarked that in signal analysis (and in Section 1.4) it is often preferable to set the integration time from $-T/2$ to $T/2$ in order to take full advantage of a number of symmetry properties. Conversely, in the analysis of system functions the lower integration limit is set at $t = 0$ because the response of the system starts at $t = 0$.

1.6 Close-in noise spectrum

The spectrum of a clock signal is broadened by noise. Experimental observation indicates that the power associated with the most interesting noise phenomena is clustered in a very narrow band around the carrier frequency. Conversely, the oscillator noise is well characterized by the low-frequency processes that describe the amplitude and the phase or frequency fluctuations. This introductory section provides a short summary of classical material available in a number of references, among which I prefer [89, 61, 102, 103].

1.6.1 Phase noise

The most frequently used tool for describing oscillator phase noise is

$S_\varphi(f)$, defined as the one-sided *power spectral density* of the random phase fluctuation $\varphi(t)$.

The physical dimension of $S_\varphi(f)$ is rad^2/Hz. Another tool of great interest is the phase-time power spectral density

$$S_x(f) = \frac{1}{(2\pi \nu_0)^2} S_\varphi(f). \tag{1.67}$$

The above relationship follows immediately from the definition of the phase time, $x(t) = \varphi(t)/(2\pi \nu_0)$. The physical dimension of $S_x(f)$ is s^2/Hz, i.e. s^3.

Manufacturers and engineers prefer the quantity $\mathscr{L}(f)$ (pronounced "script el") to $S_\varphi(f)$, where

$$\mathscr{L}(f) = \frac{1}{2} S_\varphi(f) \qquad \text{(definition of } \mathscr{L}(f)\text{)} \tag{1.68}$$

is given in dBc/Hz, i.e. the quantity used in practice is

$$10 \log_{10}\left[\tfrac{1}{2} S_\varphi(f)\right] \qquad \text{or equivalently } 10 \log_{10}\left[S_\varphi(f)\right] - 3 \text{ dB}$$

and is given in dBc/Hz. The unit dBc/Hz stands for "dB below the carrier in a 1 Hz bandwidth." The IEEE standard 1139 [103] recommends reporting the phase-noise spectra as $\mathscr{L}(f)$. Historically, $\mathscr{L}(f)$ derives from early attempts to measure the oscillator noise with a spectrum analyzer, for it was originally defined as follows:

$$\mathscr{L}(f) = \frac{\text{one-sideband noise power in 1 Hz bandwidth}}{\text{carrier power}} \qquad \text{(obsolete definition)}. \tag{1.69}$$

The definition (1.69) has been abandoned in favor of (1.68) for the following reasons.

Table 1.3 Most frequently encountered phase-noise processes

Law	Slope	Noise process	Units of b_i^a
$b_0 f^0$	0	white phase noise	rad^2/Hz
$b_{-1} f^{-1}$	-1	flicker phase noise	rad^2
$b_{-2} f^{-2}$	-2	white frequency noise (or random walk of phase)	$\text{rad}^2\,\text{Hz}$
$b_{-3} f^{-3}$	-3	flicker frequency noise	$\text{rad}^2\,\text{Hz}^2$
$b_{-4} f^{-4}$	-4	random walk of frequency	$\text{rad}^2\,\text{Hz}^3$

^{a} For brevity, convention dictates that the coefficients b_i are all given in rad^2/Hz instead of in their correct units. The unit rad^2/H_z refers to S_φ (1 Hz); see (1.70).

- The definition (1.69) does not discriminate between amplitude-modulated (AM) noise and phase-modulated (PM) noise.
- In practice, $\mathscr{L}(f)$ is always measured with a phase detector, which down-converts the carrier to dc. Hence, the definition (1.69) is experimentally incorrect because the upper and lower sidebands overlap.
- With the definition (1.69), significant discrepancies between $\mathscr{L}(f)$ and $S_\varphi(f)$ can arise in the presence of large noise.

Interestingly, the 1988 version of the IEEE standard 1139 [47] reports the obsolete definition (1.69), warning that it is to be replaced by (1.68). In the 1999 version of the same standard [103], the obsolete definition is not even mentioned.

1.6.2 Power law

A model that has been found almost indispensable in describing oscillator phase noise is the power-law function

$$S_\varphi(f) = \sum_{\substack{i=-4 \\ \text{(or less)}}}^{0} b_i f^i . \tag{1.70}$$

Phase-noise spectra are (almost) always plotted on a log–log scale, where a term f^i maps into a straight line of slope i, i.e. the slope is $i \times 10$ dB/decade. The main noise processes and their power-law characterization are listed in Table 1.3. Additionally, Table 1.4 provides a number of useful relationships discussed in Section 1.7.

In oscillators we find all the terms in (1.70) and sometimes additional terms with higher slope, while in two-port components white noise and flicker noise are the main processes. In fact, the phase noise of a two-port device cannot be steeper than $1/f$ for $f \to 0$, otherwise the group delay would diverge rapidly. This is why higher-slope phenomena in amplifiers are in fact "bumps" that appear superposed on the flicker noise. For example, this is the case for the $1/f^5$ noise of thermal origin, to be discussed in Chapter 2.

1.6.3   Frequency noise

Another way to describe oscillator noise is to ascribe its randomness to the frequency fluctuation $(\Delta\omega)(t)$ or $(\Delta\nu)(t)$ instead of to the phase (see (1.15) and (1.16)). The time-domain derivative maps into multiplication by $j\omega$ in the Fourier transform, thus into multiplication by ω^2 in the spectrum. Hence, temporarily using Ω for the Fourier frequency and ω for the carrier, it holds that $S_{\Delta\omega}(\Omega) = \Omega^2 S_\varphi(\Omega)$ and thus

$$S_{\Delta\nu}(f) = f^2 \, S_\varphi(f) \, , \tag{1.71}$$

because $\Delta\nu = (1/2\pi)\Delta\omega$, and $S_{\Delta\nu} = (1/4\pi^2)S_{\Delta\omega}$.

The use of $S_{\Delta\nu}(f)$ is common in the context of lasers, while $S_y(f)$ is more common in radio-frequency metrology. Recalling the definition of the fractional frequency fluctuation $y(t) = (\Delta\nu)(t)/\nu_0$, Eq. (1.71) becomes

$$S_y(f) = \frac{f^2}{\nu_0^2} \, S_\varphi(f) \, . \tag{1.72}$$

Of course, the power-law model also applies to $S_{\Delta\nu}(f)$ and to $S_y(f)$. In the literature the coefficients of $S_y(f)$ are denoted by h_i. Comparing (1.70) and (1.72), we find

$$S_y(f) = \sum_{i=-2}^{2} h_i f^i \tag{1.73}$$

with

$$h_i = \frac{1}{\nu_0^2} b_{i-2} \, . \tag{1.74}$$

In a comparable way to that for the power-law model of $S_\varphi(f)$, it is generally accepted that the coefficients h_i are each given in Hz^{-1}, corresponding to the physical dimension of $S_y(f)$, instead of the unit appropriate to each one.

1.6.4   Amplitude noise

The effect of amplitude noise in oscillators may be not negligible, for a variety of reasons. An example is the sensitivity of the resonant frequency to the power in the piezoelectric quartz, due to the inherent nonlinearity of the quartz lattice and known as the "isochronism defect" [42]. Another example is the effect of radiation pressure in cryogenic sapphire resonators [18], which turns the amplitude noise into frequency noise.

A power law is suitable for describing the amplitude noise of the clock signal and that of a two-port component as well. In all cases, the dominant phenomena are the white (f^0) noise and the flicker ($1/f$) noise. Bumps can be present in some portions of the frequency axis, but the asymptotic law cannot be steeper than $1/f$ at low frequencies otherwise the signal amplitude would diverge rapidly.

A detailed discussion about AM noise is available in reference [79].

1.7 Time-domain variances

In this section we use the three types of average introduced in Section 1.3, namely: the time-domain average, denoted by an overbar on the variable, as in $\overline{y}$; the average of N values, denoted by $\langle\;\rangle_N$; and the mathematical expectation, denoted by $\mathbb{E}\{\;\}$. Some care is necessary to avoid confusion.

The reading $\overline{v}_k(\tau)$ of a frequency counter[9] that measures the frequency $v(t)$ on a time interval of duration τ starting at time $k\tau$ represents the time average of $v(t)$:

$$\overline{v}_k(\tau) = \frac{1}{\tau}\int_{k\tau}^{(k+1)\tau} v(t)\,dt \;. \tag{1.75}$$

This is converted into $\overline{y}_k(\tau)$ using the normalization $\overline{y}_k = (\overline{v}_k - v_0)/v_0$, giving

$$\overline{y}_k(\tau) = \frac{1}{\tau}\int_{k\tau}^{(k+1)\tau} y(t)\,dt \;. \tag{1.76}$$

In practice, the resolution of a frequency counter is often insufficient. When this is the case, it is convenient to measure a low-frequency beat note v_b obtained by down-converting the main frequency v_0 with an appropriate reference. The resolution improves by a factor v_0/v_b. This experimental trick does not affect the development below.

Given N samples $\overline{y}_k(\tau)$, the classical variance is

$$\sigma_y^2 = \frac{1}{N-1}\sum_{k=1}^{N}\left[\overline{y}_k - \langle\overline{y}\rangle_N\right]^2 \quad \text{(VAR)}\;. \tag{1.77}$$

In the presence of flicker, random walk, and other phenomena diverging at low frequencies, the classical variance VAR depends on N and on τ, which remain to be specified. It also has a bias that grows with large measurement times. In the quest for a less biased estimator, the Allan variance and the modified Allan variance have been developed.

1.7.1 Allan variance (AVAR)

The most often used tool for the time-domain characterization of oscillators is the Allan variance (or the Allan deviation). The Allan variance

$$\sigma_y^2(\tau) = \mathbb{E}\left\{\tfrac{1}{2}\left[\overline{y}_{k+1} - \overline{y}_k\right]^2\right\} \quad \text{(AVAR)} \tag{1.78}$$

is defined as the expectation of the two-sample variance, i.e. the classical variance (1.77) evaluated for $N = 2$. The deviation is the square root of the variance:

$$\sigma_y(\tau) = \sqrt{\sigma_y^2(\tau)} = \sqrt{\text{AVAR}} \quad \text{(ADEV)}\;. \tag{1.79}$$

It is assumed that the two samples are contiguous in time. If the samples $\overline{y}_k(\tau)$ are not contiguous then a correction is necessary, which depends on the noise type [3].

---

[9] The reader should be warned that some frequency counters use a triangular weight function [80, 25], instead of the uniform weight function needed here.

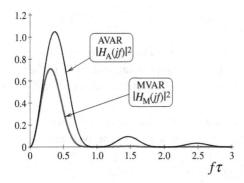

Figure 1.6 Transfer functions $|H_A(jf)|^2$ and $|H_M(jf)|^2$ of the Allan variances.

In practice, the statistical expectation is replaced by the simple mean. Given a stream of M contiguous samples $\bar{y}_k(\tau)$, we have $M-1$ differences $\bar{y}_{k+1} - \bar{y}_k$. Thus the measured Allan variance is

$$\sigma_y^2(\tau) = \frac{1}{2(M-1)} \sum_{k=1}^{M-1} (\bar{y}_{k+1} - \bar{y}_k)^2 \ . \tag{1.80}$$

Frequency-domain interpretation

The Allan variance (1.78) is equivalent to a filter or transfer function $H_A(jf)$, see Fig. 1.6, applied to $S_y(f)$:

$$\sigma_y^2(\tau) = \int_0^\infty S_y^l(f) |H_A(jf)|^2 \, df \ , \tag{1.81}$$

$$|H_A(jf)|^2 = 2 \frac{\sin^4 \pi\tau f}{(\pi\tau f)^2} \ . \tag{1.82}$$

This filter has an equivalent noise bandwidth of half an octave, having a main lobe with a peak at $f\tau \approx 0.37$. Consequently,

$$\int_0^\infty |H_A(jf)|^2 \, df = \frac{1}{2\tau} \ ; \tag{1.83}$$

thus the filter response to a white noise $S_y(f) = h_0$ is $\sigma_y^2(\tau) = h_0/(2\tau)$.

Wavelet interpretation

Writing $\bar{y}_k(\tau)$ as the integral (1.76), the Allan variance can be given as

$$\sigma_y^2(\tau) = \mathbb{E}\left\{ \left[\int_{-\infty}^{+\infty} y(t) \, w_A(t) \, dt \right]^2 \right\} , \tag{1.84}$$

AVAR

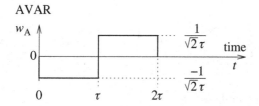

MVAR

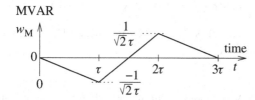

Figure 1.7 Weight function of the Allan variances $\sigma_y^2(\tau)$ and mod $\sigma_y^2(\tau)$. Reproduced from [80] with the permission of the AIP, 2007.

where the weight function (Fig. 1.7) is

$$
w_A = \begin{cases}
-\dfrac{1}{\sqrt{2}\tau} & 0 < t < \tau \,, \\[2mm]
\dfrac{1}{\sqrt{2}\tau} & \tau < t < 2\tau \,, \\[2mm]
0 & \text{elsewhere} \,.
\end{cases}
\tag{1.85}
$$

This is similar to a wavelet variance; however, the weight function $w_A(t)$ differs from that for a Haar wavelet in that it is normalized for power instead of energy. In fact, the energy[10] of w_A,

$$
E\{w_A\} = \int_{-\infty}^{+\infty} w_A^2(t)\,dt = \frac{1}{\tau} \,,
\tag{1.86}
$$

is proportional to $1/\tau$, while the energy of a wavelet is equal to 1, by definition.

Interestingly, in (1.84) $\sigma_y^2(\tau)$ has the structure of a squared scalar product. The weight function $w_A(t)$ is odd with respect to $t = \tau$ (Fig. 1.7), and thus the Allan variance AVAR detects only the odd part of the input signal. However, the noise power is equally divided into odd functions and even functions. Consequently, AVAR detects half the noise power. This is related to the fact that $E\{w_A\} = 1/\tau$, while the response to the white noise $S_y(f) = h_0$ is $\sigma_y^2(\tau) = h_0/(2\tau)$.

1.7.2 Modified Allan variance (MVAR)

The modified Allan variance MVAR [96] (see Fig. 1.7) was introduced to improve the "behavior" of AVAR in the presence of fast noise processes. It is defined, for $\tau = n\tau_0$, n

---

[10] Applying statistics to physics, the concepts of power and energy inevitably relate to physical quantities, expressed in joules or watts or dimensionless mathematical entities, as in (1.86).

an integer, as

$$\mathrm{mod}\,\sigma_y^2(\tau) = \mathbb{E}\left\{\frac{1}{2}\left[\frac{1}{n}\sum_{i=0}^{n-1}\left(\frac{1}{\tau}\int_{(i+n)\tau_0}^{(i+2n)\tau_0}y(t)\,dt - \frac{1}{\tau}\int_{i\tau_0}^{(i+n)\tau_0}y(t)\,dt\right)\right]^2\right\}\quad\text{(MVAR)}.$$

$$(1.87)$$

The deviation is the square root of the variance:

$$\mathrm{mod}\,\sigma_y(\tau) = \sqrt{\mathrm{mod}\,\sigma_y^2(\tau)} = \sqrt{\mathrm{MVAR}}\qquad\text{(MDEV)}.\qquad (1.88)$$

Once again, the statistical expectation is replaced by the mean of contiguous values, trusting the ergodicity of the process.

Frequency-domain interpretation

In the frequency domain, the MVAR computation is equivalent to the application of a filter or transfer function $H_M(jf)$ applied to $S_y(f)$:

$$\mathrm{mod}\,\sigma_y^2(\tau) = \int_0^\infty S_y^I(f)\,|H_M(jf)|^2\,df\,,\qquad (1.89)$$

whose shape depends on n. On increasing n, the shape converges rapidly to

$$\lim_{\substack{n\to\infty\\ n\tau_0=\tau}} |H_M(jf)|^2 = 2\,\frac{\sin^6\pi\tau f}{(\pi\tau f)^4}\qquad\text{(Fig. 1.6)}.\qquad (1.90)$$

For $n \gtrsim 10$, the asymptotic form is accurate enough for most practical purposes. This filter has an equivalent noise bandwidth of a quarter octave, with a peak at $f\tau \approx 0.31$ [13].

Wavelet interpretation

The modified Allan variance can also be interpreted as a wavelet variance

$$\mathrm{mod}\,\sigma_y^2(\tau) = \mathbb{E}\left\{\left[\int_{-\infty}^{+\infty}y(t)w_M(t)\,dt\right]^2\right\},\qquad (1.91)$$

in which the weight function converges to

$$w_M = \begin{cases} -\dfrac{1}{\sqrt{2\tau^2}}t & 0 < t < \tau\,,\\[2mm] \dfrac{1}{\sqrt{2\tau^2}}(2t-3\tau) & \tau < t < 2\tau\,,\\[2mm] -\dfrac{1}{\sqrt{2\tau^2}}(t-3\tau) & 2\tau < t < 3\tau\,,\\[2mm] 0 & \text{elsewhere} \end{cases}\qquad (1.92)$$

for $\tau_0 \ll \tau$, or equivalently for $n \gg 1$ (see Fig. 1.7). Once again, the weight function differs from a wavelet in that it is normalized for power instead of energy. In fact, the energy of the weight function,

$$E\{w_M\} = \int_{-\infty}^{+\infty}w_M^2(t)\,dt = \frac{1}{2\tau}\,,\qquad (1.93)$$

is proportional to $1/\tau$. Interestingly, $w_M(t)$ is an odd function with respect to $t = 3\tau/2$, and thus MVAR detects only half the noise power. The same fact was observed in the case of AVAR.

1.7.3   AVAR versus MVAR

The modified Allan variance improves on the Allan variance in that it discriminates between white PM noise and flicker PM noise. However, for a given measurement time τ the measurement of $\mathrm{mod}\,\sigma_y(\tau)$ takes longer than $\sigma_y(\tau)$. The reason is that the weight function in (1.78) takes a time 2τ and the weight function in (1.87) takes a time 3τ.

A problem with both variances is that, for the same noise process, $\sigma_y^2(\tau)$ differs from $\mathrm{mod}\,\sigma_y^2(\tau)$. In general $\mathrm{mod}\,\sigma_y^2(\tau) < \sigma_y^2(\tau)$; the actual relationship depends on the noise process. This goes with two facts,

$$E\{w_M\} = \frac{1}{2}E\{w_A\} \tag{1.94}$$

and

$$\int_0^\infty |H_M(jf)|^2\,df = \frac{1}{2}\int_0^\infty |H_A(jf)|^2\,df. \tag{1.95}$$

Thus, the AVAR response to a white frequency noise $S_y(f) = h_0$ is $\sigma_y^2(\tau) = h_0/(2\tau)$, while the response of MVAR to the same noise is $\mathrm{mod}\,\sigma_y^2(\tau) = h_0/(4\tau)$.

1.8   Relationship between spectra and variances

Figure 1.8 and Table 1.4 provide a summary of the relationship between the spectra and the Allan variance. The relationship between $S_\varphi(f)$ and $S_y(f)$ is exact, thus it is always possible to convert between $S_\varphi(f)$ and $S_y(f)$ in both directions without loss of information.

In contrast, the conversion from spectra to variances is always approximate because we cannot evaluate the integrals (1.81) and (1.89) completely. Moreover, the conversion from $\sigma_y^2(\tau)$ to $S_y(f)$ is not free from errors in the general case [39]. This is related to the presence of side lobes at $f\tau \approx 1.5, 2.5, \ldots$ in the band-pass filter $|H_A(jf)|^2$ (see (1.82) and Fig. 1.6), which mix up the spectrum. A similar phenomenon occurs in the modified Allan variance (1.87).

A further problem is that the variances detect only the components of the signal that have odd symmetry, with respect to $t = \tau$ in the case of AVAR and to $t = 3\tau/2$ in the case of MVAR. This blindness to even functions of time makes the variance phase sensitive when the signal is a deterministic modulation, such as the 50–60 Hz mains supply. Of course, spectral analysis has no preference for odd or even signals.

That said, it is clearly unfortunate that $\sigma_y(\tau)$ has been adopted as the standard for the characterization of clocks rather than $S_y(f)$. However, $\sigma_y(\tau)$ and $\mathrm{mod}\,\sigma_y(\tau)$ incorporate

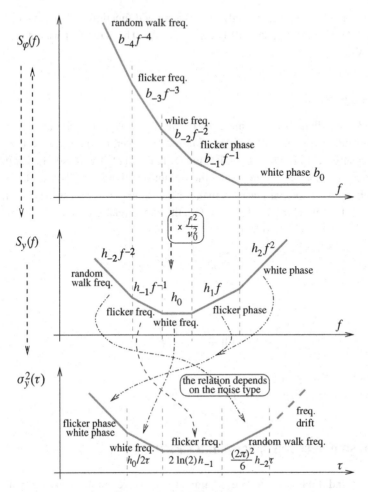

Figure 1.8 Power laws for spectra and Allan variance.

the effects of spurs, which are difficult to evaluate from the spectra. Finally, $\sigma_y(\tau)$ was easier to measure with the early instruments.

1.9 Experimental techniques

1.9.1 Spectrum analyzer

When used to measure a quasi-perfect sinusoid like the clock signal (1.6), the traditional spectrum analyzer suffers from the following limitations.

1. The IF bandwidth is too large to resolve the noise phenomena.
2. The instrument is unable to discriminate between AM noise and PM noise.

Table 1.4 Noise types, power spectral densities, and Allan variance

Noise type	$S_\varphi(f)$	$S_y(f)$	$S_\varphi \leftrightarrow S_y$	$\sigma_y^2(\tau)^a$	$\mathrm{mod}\,\sigma_y^2(\tau)^a$
white PM	b_0	$h_2 f^2$	$h_2 = \dfrac{b_0}{\nu_0^2}$	$\dfrac{3 f_H h_2}{(2\pi)^2}\,\tau^{-2},$ $2\pi\tau f_H \gg 1$	$\dfrac{3 f_H \tau_0 h_2}{(2\pi)^2}\,\tau^{-3}$
flicker PM	$b_{-1} f^{-1}$	$h_1 f$	$h_1 = \dfrac{b_{-1}}{\nu_0^2}$	$[1.038 + 3\ln(2\pi f_H\tau)]$ $\times \dfrac{h_1}{(2\pi)^2}\,\tau^{-2}$	$0.084\,h_1\tau^{-2},$ $n \gg 1$
white FM	$b_{-2} f^{-2}$	h_0	$h_0 = \dfrac{b_{-2}}{\nu_0^2}$	$\dfrac{1}{2}h_0\,\tau^{-1}$	$\dfrac{1}{4}h_0\,\tau^{-1}$
flicker FM	$b_{-3} f^{-3}$	$h_{-1} f^{-1}$	$h_{-1} = \dfrac{b_{-3}}{\nu_0^2}$	$2\ln(2)\,h_{-1}$	$\dfrac{27}{20}\ln(2)\,h_{-1}$
random walk FM	$b_{-4} f^{-4}$	$h_{-2} f^{-2}$	$h_{-2} = \dfrac{b_{-4}}{\nu_0^2}$	$\dfrac{(2\pi)^2}{6}h_{-2}\tau$	$0.824\dfrac{(2\pi)^2}{6}h_{-2}\tau$
linear frequency drift $\dot{y}$				$\dfrac{1}{2}(\dot{y})^2\,\tau^2$	$\dfrac{1}{2}(\dot{y})^2\,\tau^2$

^{a} f_H is the low-pass cutoff frequency, needed for the noise power to remain finite.

3. The dynamic range is insufficient for the instrument to detect the noise sidebands; it is "dazzled" by the strong carrier.
4. The noise and the frequency fluctuations of the local oscillator are larger than the noise of most good oscillators of interest.

The first two problems are solved by smart digital technology in some modern analyzers, which are capable of phase-noise measurement. However, the dynamic range is still limited by the available technology of frequency conversion. The higher dynamic range requires a carrier cancellation method, such as direct phase measurement with a double balanced mixer or a bridge, which cannot be implemented[11] by a spectrum analyzer. Finally, the spectrum analyzer scans the input range with a voltage-controlled oscillator. Owing to the wide tuning range, the spectral purity of such an oscillator is insufficient to measure the phase noise of most fixed-frequency oscillators. In conclusion, even if some modern spectrum analyzers are capable of measuring phase noise, this function is generally inadequate in numerous practical cases.

1.9.2 Phase-locked loop

The scheme on which virtually all commercial phase-noise measurement systems are based includes a phase-locked loop (PLL), as shown in Fig. 1.9. The double-balanced mixer is saturated at both inputs by two signals in quadrature, and it convertsthe phase

[11] Actually, a carrier suppression technique was proposed as an extension of a spectrum analyzer to measure $\mathscr{L}(f)$ [50]. After the prototype, this solution seems not to have been used further.

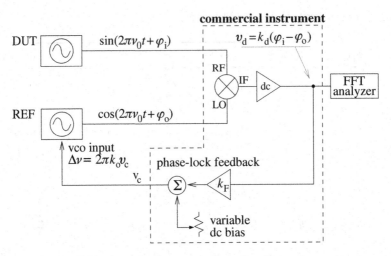

Figure 1.9 Practical scheme for the measurement of oscillator phase noise.

difference $\varphi_i - \varphi_o$ into a voltage $v_d = k_d(\varphi_i - \varphi_o)$, fed back to the VCO input and detected by the FFT analyzer. In this configuration, the PLL is a high-pass filter for the oscillator noise. In fact, the noise measurement relies on the fact that the locked oscillator does not track the fast fluctuations of the oscillator under test, so that the error signal is available for the noise measurement. The PLL ensures the quadrature relationship for the duration of the measurement. The reference oscillator is pulled to the DUT frequency v_0 by the addition of a dc bias to the control voltage.

Using Laplace transforms, denoted by the upper case letter corresponding to the time-domain signal, the PLL transfer function is

$$\frac{V_d(s)}{\Phi_i(s)} = \frac{k_d\, s}{s + k_L}, \tag{1.96}$$

where

$$k_L = k_o k_d k_F \qquad \text{(loop gain)} \tag{1.97}$$

is the loop gain. The VCO gain k_o is given in rad/(V s). The mixer phase-to-voltage gain k_d (V/rad) also accounts for the gain of the dc amplifier, and k_F is the gain of the feedback-path amplifier. In order to rewrite (1.96) for the noise spectra, we take the square modulus after substituting $s = j2\pi f$ and obtain

$$\frac{S_v(f)}{S_\varphi(f)} = \frac{k_d^2\, f^2}{f^2 + k_L^2}. \tag{1.98}$$

Naively, one would be inclined to set a low loop gain k_L, so that the cutoff frequency $f_c = k_L/(2\pi)$ is low enough to leave the device under test (DUT) free to fluctuate at all the frequencies in the spectral analysis. Of course, k_L must still be high enough for the DUT to track the reference oscillator in the long term, which guarantees the quadrature relationship at the double-balanced mixer ports. With such loose PLLs, (1.98) is approximated by $S_v(f)/S_\varphi(f) = k_d^2$, from which we estimate $S_\varphi(f)$.

A more sophisticated approach consists of setting up a tight phase locking; of course, the high-pass function (1.98) must be measured accurately and subtracted from the measured spectral density. This approach has the following advantages.

1. In practice, the two oscillators are more or less loosely coupled by leakage. This gives the wrong impression of a phase noise lower in magnitude than the actual value. Of course, the leakage is hardly reproducible because it depends on the experimental layout. With an appropriate choice of the operating parameters, however, a tight loop can override the effect of leakage.
2. The phase noise rises as $1/f^3$, or as $1/f^4$ at low frequencies. After tight locking, $S_v(f)$ rises as $1/f$, or as $1/f^2$ at low frequencies. This reduces the burden of dynamic range at the FFT input.

The most appropriate time constant is determined on the basis of experimental criteria. A good starting point is the DUT cutoff frequency, where the $1/f$ or $1/f^2$ phase noise turns into $1/f^3$.

The assumption that the reference oscillator is ideally stable is realistic only in some routine measurements where a low-noise reference oscillator is available. Conversely, in the measurement of ultra-stable oscillators the PLL is used to compare two equal oscillators. In these conditions, the DUT noise is half the raw value of $S_\varphi(f)$.

The background noise is determined by the following facts.

- The white noise floor is chiefly due to the dc low-noise amplifier. The reason is the poor phase-to-voltage conversion gain of the mixer, which is about 0.1 to 0.5 V/rad. White noise values of -160 to -170 dB rad^2/Hz ($b_0 = 10^{-16}$ to 10^{-17} rad^2/Hz) are often seen in practice.
- The flicker noise is due to the mixer. Common values are -120 dB rad^2/Hz at 1 Hz offset ($b_{-1} = 10^{-12}$ rad^2/Hz) for microwave mixers and -140 dB rad^2/Hz ($b_{-1} = 10^{-14}$ rad^2/Hz) for RF units up to about 1 GHz.
- In some cases, the amplitude noise pollutes the phase-noise measurement through the mixer power-to-dc-offset conversion [81]. This is due to symmetry defects in real mixers.

Exercises

1.1   A cosine signal has frequency 10 MHz and rms amplitude 2 V. Write the clock signal in the form (1.6) and sketch the associated phasor.

1.2   Sketch the complex envelope (phasor) associated with the signal of the previous exercise, after adding a phase noise of 2×10^{-4} rad rms. Repeat for an amplitude noise of 5×10^{-4} but no phase noise.

1.3 The 10 MHz signal of the previous exercise, as affected by a phase noise equal to 2×10^{-4} rad rms, is multiplied to 100 MHz and to 200 MHz using noise-free electronic circuits. Sketch the complex envelopes (phasors) associated with the 100 MHz and 200 MHz outputs.

1.4 Your wristwatch lags 2 seconds per day. Plot $x(t)$ and $y(t)$.

1.5 As in the previous exercise, your wrist watch lags 2 seconds per day. Plot $\varphi(t)$ and $(\Delta v)(t)$ assuming that the internal oscillator is a 2^{15} Hz quartz.

1.6 Repeat the previous exercise replacing the quartz by a swinging wheel oscillating at 5 Hz.

1.7 Your wrist watch is free from noise and errors. However, reading the time on it takes in a random error of 0.5 s max. Plot the Allan deviation $\sigma_y(\tau)$. Assume that the error distribution is uniform in $(-500, +500)$ ms, and that the spectrum is white.

1.8 A near-dc amplifier has gain 20 dB, white noise 4 nV/$\sqrt{\text{Hz}}$, and flicker noise 4 nV/$\sqrt{\text{Hz}}$ at 1 Hz. Plot the noise spectrum of the input and output voltages.

1.9 Plot the Allan deviations of the input and output voltages for the near-dc amplifier in the previous exercise. Tip: use Table 1.4, with v instead of y. Of course, the physical unit of such a σ_v is V.

1.10 Write the transfer function $H(s)$ of a simple RC low-pass filter (R in series, C to ground) with $R = 1$ kΩ and $C = 2.2$ nF, and sketch the Bode plot.

1.11 The input of the above low-pass filter is a 200 nV/$\sqrt{\text{Hz}}$ white noise source, uniform from dc to 1 GHz. Sketch the output noise spectrum and calculate the total output power. Ignore the thermal noise added by the filter.

1.12 Thermal noise is governed by the law $P = kTB$, where kT is the thermal energy at temperature T and B is the bandwidth. Such a noise is added to an ideal 1 mW carrier. Calculate $\mathscr{L}(f)$ and $S_\varphi(f)$ at the temperatures $T_0 = 290$ K, $T_{\text{liquid N}} = 77$ K, and $T_{\text{liquid He}} = 4.2$ K.

2 Phase noise in semiconductors and amplifiers

Understanding amplifier phase noise is vital to the comprehension of oscillators. There are two basic classes of noise, additive and parametric, as illustrated in Fig. 2.1. However, their behavior, and the underlying physical mechanisms, are strikingly different. This chapter focuses on amplifiers but includes a few words on the photodetector, which forms part of the resonator–amplifier loop in the emerging photonic (or optoelectronic) oscillators. The same principles apply to other electronic and optoelectronic components.

Random noise is normally present in electronic circuits. When such noise can be represented as a voltage or current source that can be added to the signal, it is called additive noise. Another mechanism can be present, referred to as parametric noise, in which a near-dc process modulates the carrier in amplitude, in phase, or in a combination thereof. The most relevant difference is that additive noise is always present, while parametric noise requires nonlinearity and the presence of a carrier. Additive noise is generally white, while parametric noise can have any shape around ν_0, depending on the near-dc phenomenon. Additive noise originates in the region around the carrier frequency ν_0 while parametric noise is brought there from near-dc by the carrier. This is made evident by a gedankenexperiment (thought experiment) in which we inspect the amplifier output around ν_0 with an ideal spectrum analyzer, switching the input carrier on and off. A noise process such as $1/f$ noise cannot be present around ν_0 when the carrier is switched off, while the additive noise will still be there unchanged.

Among parametric noises, *flicker* ($1/f$ noise) is so important that the term "parametric noise" is sometimes used synonymously with it. Parametric noise originating from near-dc white noise is generally not relevant in practice. Environmental fluctuations often show up as parametric noise at very low offset frequencies.

2.1 Fundamental noise phenomena

2.1.1 Thermal noise

The power spectral density of blackbody radiation is described by the law

$$S(\nu) = \frac{h\nu}{e^{h\nu/kT} - 1} \qquad \text{(W/Hz, or J)} \tag{2.1}$$

$$\simeq kT \qquad \text{for } h\nu \ll kT . \tag{2.2}$$

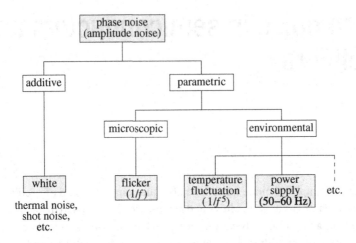

Figure 2.1 Noise types in amplifiers. The same classification is valid for both phase and amplitude noise.

Energy equipartition in classical thermodynamics states that the energy is $\frac{1}{2}kT$ per degree of freedom. Electromagnetic radiation has two degrees of freedom, electric field and the magnetic field, and so the energy of blackbody radiation is kT. Blackbody radiation is white noise at low frequencies, decreasing as $1/(e^{h\nu/kT} - 1)$, below the cutoff frequency kT/h. At the cutoff, the photon energy $h\nu$ equals the thermal energy kT. For reference, the cutoff occurs at 6.04 THz (49.6 μm wavelength) at 290 K, and at 87.5 GHz (3.42 mm wavelength) at liquid He temperature (4.2 K). The physical dimension of the thermal noise spectrum is that of energy, expressed in joules. Yet in spectral analysis we prefer the equivalent unit W/Hz.

Resistor noise, observed and analyzed theoretically in 1928 [58, 72], comes from the coupling of the blackbody radiation of the resistor to the rest of the electric circuit. The power spectrum density is often denoted by the symbol N, taken to be independent of ν, instead of $S(\nu)$:

$$N = kT \qquad \text{(resistor noise, W/Hz)} . \qquad (2.3)$$

The spectrum of the random voltage available[1] at the resistor ends in V^2/Hz is

$$S_v(\nu) = kTR \qquad \text{(resistor terminated)} \qquad (2.4)$$

if the resistor is impedance-matched to a load, and

$$S_v(\nu) = 4kTR \qquad \text{(resistor open)} \qquad (2.5)$$

if the resistor terminals are left open.

---

[1] In the jargon of electric networks, the term "available power" refers to the maximum power delivered by the generator, which occurs when the generator is impedance-matched to the load. By extension, the term "available" also applies to voltage and to current, referring to the condition of impedance matching.

Example 2.1. The available noise power of a resistor at the temperature $T = 290$ K is $kT = 4 \times 10^{-21}$ W/Hz. Assuming a resistance $R = 1$ kΩ, the spectrum of the available noise voltage is $kTR = 4 \times 10^{-18}$ V^2/Hz; thus $\sqrt{kTR} = 2$ nV/$\sqrt{\text{Hz}}$. Leaving the resistor terminals open, the spectrum is $4kTR = 1.6 \times 10^{-17}$ V^2/Hz; thus $\sqrt{4kTR} = 4$ nV/$\sqrt{\text{Hz}}$.

2.1.2 Shot noise

In most electronic devices the conduction of electricity takes place as a stream of separate electrical charges $q = 1.602 \times 10^{-19}$ C (the charge on an electron), each one giving rise to a current pulse. This is a random current $i(t)$ that originates the white noise of the spectrum,

$$S_i(v) = 2q\bar{i} \quad \text{(A}^2\text{/Hz)}, \tag{2.6}$$

where $\bar{i}$ is the average current. Equation (2.6) is independent of whether the carriers are electrons or holes or a combination of these. The power spectrum of the random current $i(t)$ flowing into a resistor R is

$$S_i(v) = 2q\bar{i}\,R \quad \text{(W/Hz)}. \tag{2.7}$$

2.1.3 Flicker noise

The existence of *excess noise*, later called *contact noise* or *flicker noise*, was initially recognized by Johnson [57]; and after that it was studied in thin films [4, 5] and carbon microphones [19]. Later, it was observed that noise with a spectrum scaling of the type $1/f^x$, with $x \in (0.8, 1.2)$, is present in an wide variety of physical phenomena, such as stellar emissions, lake turbulences, Nile flooding, and of course virtually all electronic devices. In the absence of a unifying theory that explains why flicker is so ubiquitous, we rely on models, the most accredited of which are due to Hooge [49] and to McWhorter [69].

2.2 Noise temperature and noise figure

The noise temperature and the noise figure, most often used in the context of amplifiers and radio receivers, are parameters that describe the white noise of a device (Fig. 2.2).

In analogy with the thermal noise kT, the amplifier noise spectrum N_a is referred to the amplifier input and given in terms of its *equivalent noise temperature* T_a:

$$N_a = kT_a \quad \text{(noise temperature } T_a\text{)}. \tag{2.8}$$

The amplifier is always input-terminated to some dissipative load, which contributes its thermal noise, $N_r = kT_r$. The total equivalent noise N_e referred to the input is therefore

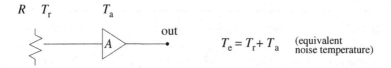

$$T_e = T_r + T_a \quad \begin{array}{l}\text{(equivalent}\\\text{noise temperature)}\end{array}$$

$$F = \frac{T_0 + T_a}{T_0} \quad \text{(noise figure)}$$

Figure 2.2 Equivalent noise temperature and noise figure in amplifiers.

given by

$$N_e = N_a + N_r \tag{2.9}$$

$$= k(T_a + T_r) \, . \tag{2.10}$$

In radio and microwave engineering, the noise parameter most commonly used by manufacturers is the *noise figure F*, defined as

$$F = \frac{(SNR)_{\text{input}}}{(SNR)_{\text{output}}} \quad \text{(definition of the noise figure } F) \, , \tag{2.11}$$

where *SNR* is the signal-to-noise ratio. In practice, the definition (2.11) is often replaced by the ratio of N_e and the thermal spectrum associated with a reference temperature T_0:

$$F = \frac{N_e}{kT_0} \quad \text{(practical noise figure } F) \, . \tag{2.12}$$

For the definition of F to be unambiguous, and for (2.12) to be equivalent to (2.11), it is necessary that the amplifier is appropriately terminated to its input impedance R_0 and that the input termination is at a standard temperature T_0. In practice, it is preferable for the entire system to be at the temperature T_0. The standard value generally adopted is

$$T_0 = 290 \, \text{K} \qquad (17\,^\circ\text{C}) \, , \tag{2.13}$$

with associated power spectrum $kT_0 = 4 \times 10^{-21}$ W/Hz, that is, -174 dB m in a 1 Hz bandwidth.

Substituting (2.12) into (2.10), with $T_r = T_0$, we find that

$$F = \frac{T_0 + T_a}{T_0} = 1 + \frac{T_a}{T_0} \, ; \tag{2.14}$$

hence

$$T_a = (F - 1)T_0 \tag{2.15}$$

and

$$N_a = (F - 1)kT_0 \, . \tag{2.16}$$

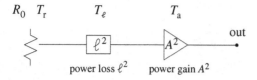

Figure 2.3 Amplifier noise in the general case of non-uniform temperature.

The practical use of the noise figure F is that the resistor thermal noise $N_r = kT$ can be replaced by the equivalent noise

$$N_e = FkT_0 . \tag{2.17}$$

The typical noise figure for low-noise amplifiers is 0.5–2 dB, depending on technology, on frequency, and on amplifier bandwidth. It is higher in low-cost amplifiers and in power amplifiers.

It is to be pointed out that the *noise temperature*, and the *noise figure* as well, account for all noise phenomena including shot noise. For example, at high optical power the equivalent noise temperature at the output of a photodetector is mainly due to shot noise, which has very little to do with the device's physical temperature or the temperature of the optical source.

Some authors describe amplitude noise and phase noise using a non-white spectrum, by introducing a frequency dependence into the noise temperature and the noise figure: thus, $T(f)$ and $F(f)$. This may be wrong in the general case, where the amounts of amplitude noise and phase noise are not the same.

2.2.1 ★ The general case of non-uniform temperature

Let us consider the general case of Fig. 2.3, in which the amplifier input is terminated to a resistor through an attenuator of power loss ℓ^2 and the whole circuit is impedance-matched to R_0. As the temperature is left arbitrary, we need to use the noise power spectral density instead of the noise figure. The noise of the termination, referred to the amplifier input, is

$$N_r = \frac{1}{\ell^2} kT_r , \tag{2.18}$$

because the attenuator attenuates the noise of the termination, as it does with any signal. The attenuator noise referred to the amplifier input is

$$N_\ell = \frac{\ell^2 - 1}{\ell^2} kT_\ell . \tag{2.19}$$

From this relationship it follows that the amplifier receives a noise kT_0 from the resistance R_0 it sees at its input when $T_r = T_\ell = T_0$.

The amplifier noise contribution is $N_a = (F - 1)kT_0$, as given by (2.10). The total noise referred to the amplifier input is $N_e = N_r + N_\ell + N_a$, i.e.

$$N_e = \frac{1}{\ell^2} kT_r + \frac{\ell^2 - 1}{\ell^2} kT_\ell + (F - 1) kT_0 . \tag{2.20}$$

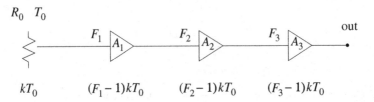

Figure 2.4 Noise temperature and noise figure in amplifiers.

The scheme of Fig. 2.3 can be generalized further by introducing impedance mismatching. The analysis is omitted here.

A non-uniform temperature is found, for example, in cryogenic oscillators and in oven-controlled quartz crystal oscillators (OCXOs). In the first case, the resonator operates at low temperatures (4–80 K), where it exhibits a high Q factor. In the case of an OCXO, the resonator is kept at the turning point of the frequency–temperature characteristics, which occurs at some 75–80 °C. This choice makes the temperature control simple and effective. In both cases, the sustaining amplifier is at a convenient temperature.

2.2.2   The noise figure of cascaded amplifiers

In a cascade of amplifiers, it is convenient to refer the total noise to the input of the first amplifier. With reference to Fig. 2.4, adding the individual contributions we divide the noise $(F_2 - 1)kT_0$ of the second amplifier by the power gain A_1^2 of the first amplifier, the noise $(F_3 - 1)kT_0$ of the third amplifier by the power gain $A_1^2 A_2^2$ of the two preceding amplifiers, etc. The final result is

$$N_e = kT_0 + (F_1 - 1)kT_0 + \frac{(F_2 - 1)kT_0}{A_1^2} + \frac{(F_3 - 1)kT_0}{A_1^2 A_2^2} + \cdots \qquad (2.21)$$

Generalizing, we add the noise $(F_m - 1)kT_0$ of the mth amplifier after dividing it by the power gain $\prod_{i=1}^{m-1} A_i^2$ of the preceding $m - 1$ amplifiers. Using the noise figure instead of the noise power, (2.21) becomes

$$F = F_1 + \frac{F_2 - 1}{A_1^2} + \frac{F_3 - 1}{A_1^2 A_2^2} + \cdots \qquad \text{(Friis formula)} , \qquad (2.22)$$

which is known as the Friis formula [33]. The important conclusion that can be drawn from (2.22) is that there is a strong dominance, in general, of the noise introduced by the first amplifier over the other noise contributions. Therefore, special care should be taken in the choice of the first stage of amplification.

2.2.3   Noise in amplifiers and photodetectors

Bipolar transistors

In modern transistors, it turns out that the thermal noise of the distributed base resistance is the most relevant noise type [100, Section 9.4]. This resistance is usually referred to as $R_{bb'}$ (also $r_{bb'}$ in some textbooks) in the pioneering hybrid-π model proposed by Giacoletto [37] and later in the popular Gummel–Poon model [44]. For the sake of completeness, the other basic noise phenomena are shot noise in the base region and in the collector and partition noise.

Field-effect transistors

The most relevant source of white noise in these devices is the channel noise fed back to the gate through the gate-channel capacitance and amplified [100, Section 9.3]. This mechanism also accounts for the increased noise figure beyond a cutoff frequency. Mosfets have a low noise temperature, generally of the order of 10 K, and capacitive input impedance. A complex network is needed to match the gate to the standard resistive input (50 Ω). This network introduces loss and thus increases the noise figure. Needless to say, a large-bandwidth design results in a high input loss, leading to a high noise figure.

High-speed solid-state photodetector

This type of photodetector is a reverse-biased junction in which the light originates a photocurrent. A detected photon creates an electron–hole pair, $+q$ and $-q$; these are accelerated in opposite directions by the electric field. Thanks to the Ramo theorem, the shape of the current pulses at the detector ends is such that the total charge transferred is q. Given that the optical power is $P(t)$ at frequency ν, the number of photons per second must be $P/(h\nu)$. The photocurrent is

$$i(t) = \frac{q\eta}{h\nu} P(t) \,, \tag{2.23}$$

where η is the quantum efficiency, i.e. the probability that the photon is detected. The quantity

$$\rho = \frac{q\eta}{h\nu} \quad \text{(responsivity, A/W)} \tag{2.24}$$

is called the responsivity. For a typical quantum efficiency $\eta = 0.8$, the responsivity ρ is about 1 A/W at $\lambda = 1.55\,\mu$m and about 0.85 A/W at $\lambda = 1.32\,\mu$m. Combining (2.6) and (2.23), the power spectral density of the detector shot noise is

$$S_i(\nu) = 2\frac{q^2\eta}{h\nu} \overline{P} \quad (A^2/Hz) \,. \tag{2.25}$$

Example 2.2. A beam of power $P = 1\,$mW at wavelength $\lambda = 1.55\,\mu$m carries 7.8×10^{15} photons per second. Assuming a quantum efficiency $\eta = 0.8$, the photocurrent $i = 1\,$mA. The shot noise $S_i = 3.2 \times 10^{-22}\,A^2/Hz$ and thus $\sqrt{S_i} = 17.9\,$pA$/\sqrt{Hz}$. When the photocurrent flows into a 50 Ω resistor, the noise is $1.6 \times 10^{-20}\,$W/Hz or 8.01×10^{-19} V^2/Hz (corresponding to $0.895\,$nV$/\sqrt{Hz}$).

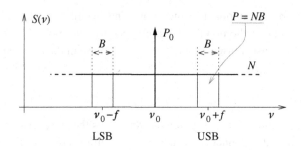

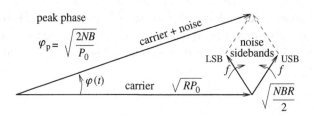

Figure 2.5 The contribution of additive noise to the phase noise.

2.3   Phase noise and amplitude noise

2.3.1   White noise

All noise processes with white-spectrum statistics are grouped under the label of *additive noise*, i.e. a random white process added to the useful signal. Of course, the noise is not correlated with the signal, nor connected with it by some conformal transformation. Under these conditions, the noise power is equally partitioned into the two degrees of freedom, that is, the in-phase and quadrature components with respect to the carrier or, equivalently, the amplitude noise and phase noise. In the case of thermal noise, we find an energy equal to $\frac{1}{2}kT$ for each degree of freedom.

We will derive the expression for the phase noise from the noise power spectral density N using the phase-vector representation of a sinusoid (Fig. 2.5). The carrier power P_0 corresponds to a vector of rms voltage $\sqrt{RP_0}$ on the real axis. The noise power in the bandwidth B is NB and thus $\frac{1}{2}NB$ in each degree of freedom, i.e. amplitude and phase. Accordingly, the phase-noise upper sideband (USB) is a random sinusoid of frequency $\nu_0 + f$ and of rms voltage

$$V_{\text{rms}} = \sqrt{\frac{1}{2}NBR} \,. \tag{2.26}$$

This is represented as a vector of rms voltage V_{rms} rotating at frequency f around the tip of the carrier vector, in quadrature with the carrier at time $t = 0$. Similarly, the phase-noise lower sideband (LSB) is a vector of the same rms voltage, rotating at frequency $-f$, and also in quadrature with the carrier at the time $t = 0$. The sum of the carrier and

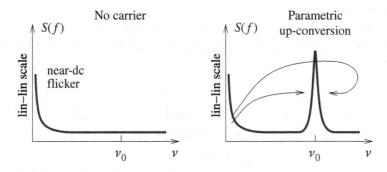

Figure 2.6 Parametric up-conversion of near-dc flicker in amplifiers. Reproduced from [87] and used with the permission of the IEEE, 2007.

the two phase-noise sidebands is a vector of peak phase

$$\varphi_p = \sqrt{\frac{2NB}{P_0}} \qquad \text{for } \varphi_p \ll 1 \tag{2.27}$$

oscillating sinusoidally around the real axis. The rms phase is

$$\varphi_{rms} = \sqrt{\frac{NB}{P_0}}. \tag{2.28}$$

The power spectral density is the average square phase noise per unit bandwidth and is independent of frequency ($S_\varphi(f) = b_0$), because this is white noise. Hence

$$b_0 = \frac{NB}{P_0} \qquad \text{(white phase noise)}, \tag{2.29}$$

which in the case of the phase noise of an amplifier can be rewritten as

$$b_0 = \frac{FkT_0}{P_0} \qquad \text{(amplifier)}. \tag{2.30}$$

2.3.2 Flicker noise

Understanding flicker noise starts from the simple observation that the output spectrum is of the white type – flat in a wide frequency range – when the carrier power is zero and that the close-in noise shows up only when the carrier power present at the amplifier output is sufficiently large (Fig. 2.6). At closer sight, it is observed that the noise sidebands grow in proportion to the carrier. The obvious consequence is that the close-in flickering results from the near-dc flicker noise that modulates the carrier in amplitude and phase. Two simple models are presented below.

Nonlinear mechanism

The first mechanism of noise up-conversion relies upon the amplifier's non-linearity, which for our purposes can be represented as a polynomial truncated at the second degree:

$$v_o(t) = a_1 v_i(t) + a_2 v_i^2(t) + \cdots \qquad (2.31)$$

The coefficient a_1 is the voltage gain denoted by A elsewhere in this book. Given a carrier $V_i e^{j\omega_0 t}$ at the input, the amplifier's noisy input signal is

$$v_i(t) = V_i e^{j\omega_0 t} + n(t) \qquad (2.32)$$

$$= V_i e^{j\omega_0 t} + n'(t) + jn''(t), \qquad (2.33)$$

where $n(t) = n'(t) + jn''(t)$ is the internally generated near-dc noise. The noise $n(t)$ is the amplifier's internal noise referred to the input and accounts for the efficiency of the modulation process. One could object that the use of the analytic signal $e^{j\omega_0 t}$ in nonlinear circuits is not allowed in the general case because it describes only the positive frequencies; the consequence is that the frequency-difference terms do not appear in the equations. Nonetheless, the use of $e^{j\omega_0 t}$ is correct in this case because we select only the up-conversion term, which brings $n(t)$ close to the carrier. Thus, combining (2.31) and (2.32) and discarding the terms far from v_0, we get the output signal:

$$v_o(t) = V_i \{ a_1 + 2a_2 [n'(t) + jn''(t)] \} e^{j\omega_0 t}. \qquad (2.34)$$

In this representation, the term $V_i a_1 e^{j\omega_0 t}$ is the output carrier and the term $V_i[2a_2 n'(t) + j2a_2 n''(t)]e^{j\omega_0 t}$ is the close-in noise associated with the output signal. The latter originates from the cross-product terms coming from the expansion of $a_2 v_i^2(t)$ in (2.31). It follows that, using the signal representation (1.6), the random amplitude and phase fluctuations are

$$\alpha(t) = 2 \frac{a_2}{a_1} n'(t), \qquad (2.35)$$

$$\varphi(t) = 2 \frac{a_2}{a_1} n''(t). \qquad (2.36)$$

Interestingly, $\alpha(t)$ and $\varphi(t)$ are independent of the input amplitude V_i. Therefore the power spectral densities $S_\alpha(f)$ and $S_\varphi(f)$ are expected to be independent of the carrier power.

According to this model, the $1/f$ noise arises from the amplifier's nonlinearity, and it should thus be possible to reduce it by an improvement in linearity. Yet, most of the linearization methods known in the literature fail in this case because they act on the entire amplifier, while the $1/f$ noise is generated at a microscopic level inside the electronic device. An example is the case of the push–pull amplifier. Because the even harmonics are canceled by symmetry, one may expect a significant reduction in the $1/f$ noise, but this does not happen because the $1/f$ noise is generated locally by the two transistors of the push–pull, and it is added statistically. Nonetheless a reduction of 3 dB in the $1/f$ noise is expected, for reasons detailed below in subsection 2.5.4. Some methods useful in reducing the $1/f$ phase noise are explained in Section 2.5.

Quasi-linear parametric mechanism

Another mechanism of near-dc noise up-conversion is the fluctuation in the amplifier gain. This is modeled by replacing the constant voltage gain A by

$$A(t) = A_0\left[1 + n(t)\right] \qquad \text{(amplifier gain)} \tag{2.37}$$

$$= A_0\left[1 + n'(t) + jn''(t)\right], \tag{2.38}$$

where $n(t) = n'(t) + jn''(t)$ is the near-dc noise. It should be noted that the noise $n(t)$ introduced here is not the same physical quantity $n(t)$ as that in (2.32), (2.33). For example, in the case of a bipolar transistor, $n''(t)$ is easily identified as a signal proportional to the fluctuating voltage that modulates the collector-base reverse voltage and in turn the capacitance. The latter affects the amplifier phase lag, which generates phase noise. Nonetheless, we will keep the analysis at a higher level of abstraction, leaving the nature of $n(t)$ unspecified.

In the presence of a carrier $V_i e^{j\omega_0 t}$ at the input, the amplifier output is

$$v_o(t) = A_0\left[1 + n'(t) + jn''(t)\right] V_i e^{j\omega_0 t}, \tag{2.39}$$

from which it follows that

$$\alpha(t) = n'(t), \tag{2.40}$$

$$\varphi(t) = n''(t). \tag{2.41}$$

Once again, $\alpha(t)$ and $\varphi(t)$ are independent of the input amplitude V_i and thus $S_\alpha(f)$ and $S_\varphi(f)$ are expected to be independent of the carrier power.

It is to be pointed out that the parametric model is *nearly linear*. In mathematics, a function $f(x)$ is linear if it exhibits the following two properties:

$$f(ax) = af(x), \tag{2.42}$$

$$f(x + y) = f(x) + f(y). \tag{2.43}$$

One can verify the property (2.42) by replacing $V_i e^{j\omega_0 t}$ by $aV_i e^{j\omega_0 t}$ in (2.39) and the property (2.43) by replacing $V_i e^{j\omega_0 t}$ by $V_1 e^{j\omega_1 t} + V_2 e^{j\omega_2 t}$ in (2.39). In both cases the deviation from linearity consists of terms containing $n'(t)$ and $n''(t)$, which are proportionally small in low-noise conditions, i.e. for $|n'(t)| \ll 1$ and $|n''(t)| \ll 1$. This is radically different from the model (2.34), where the strong nonlinearity generates harmonics of the carrier and beat notes.

Flicker noise in real amplifiers

It turns out that both mechanisms are present in real electronic devices, i.e. both nonlinearity and modulation of the gain, and both originate $1/f$ noise. In a deeper analysis, the difference between these two phenomena is more subtle than that presented here because the nonlinear approach hides a parametric modulation. As a first approximation, we can assume that in both models the statistical properties of the near-dc processes $n'(t)$ and $n''(t)$ are not affected by the carrier power. This is confirmed by the experimental observation that the amplifier phase noise given in $\mathrm{rad}^2/\mathrm{Hz}$ is roughly independent of the power in a wide range [45, 105, 46]. Some dependence on P_0 may remain in practice.

Table 2.1 Typical phase flickering of amplifiers in dB rad^2/Hz

Rating	GaAs HBT (microwave)	SiGe HBT (microwave)	Si bipolar (HF–UHF)
fair	-100		-120
good	-110	-120	-130
best	-120	-130	-140

We ascribe this to terms of order higher than v_i^2 in (2.31) and to the fact of operation in a large-signal regime, which affects the dc-bias and in turn $n'(t)$ and $n''(t)$.

In conclusion, the $1/f$ phase noise is best described by

$$S_\varphi(f) = b_{-1}f^{-1} \qquad (b_{-1} \approx \text{constant vs. } V_i),\tag{2.44}$$

where the coefficient b_{-1} is an experimental parameter of the specific amplifier. The same holds for the $1/f$ amplitude noise. Table 2.1 shows the typical flicker phase noise of commercial amplifiers. Phase flickering is related to the physical size of the amplifier's active region and to the amplifier gain. The latter is a side effect of the number of stages needed for a given gain. More details are given in subsection 2.5.3.

2.3.3 Power spectral density and corner frequency

Let us combine the white and flicker noise, under the assumption that they are independent. This is in agreement with observations on actual amplifiers. Thus we have for $S_\varphi(f)$ (and see (2.30), (2.44))

$$S_\varphi(f) = b_0 + b_{-1}\frac{1}{f},\tag{2.45}$$

$$b_0 = \frac{FkT}{P_0},$$

$$b_{-1} = \text{constant}.$$

Figure 2.7 shows the phase-noise spectrum of an amplifier. In this figure, we observe that the corner frequency, where the white noise equals the flicker noise, is

$$f_c = \frac{b_{-1}}{b_0}.\tag{2.46}$$

Recalling that $b_0 = FkT/P_0$, the above becomes

$$f_c = \frac{b_{-1}}{FkT_0}P_0.\tag{2.47}$$

As a trivial yet relevant conclusion, f_c is proportional to the carrier power P_0.

Experimental evidence

Figures 2.8 and 2.9 show the phase-noise spectra of two microwave amplifiers measured at different power levels below compression. The first amplifier (Fig. 2.8), which was

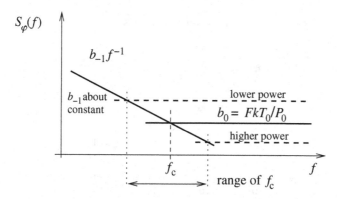

Figure 2.7 Typical phase noise of an amplifier. Reproduced from [83] with the permission of the IEEE, 2007.

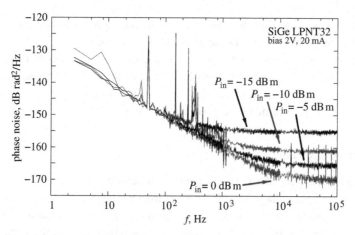

Figure 2.8 Example of amplifier phase noise measured at various input power levels below saturation. Courtesy of Rodolphe Boudot [11].

measured at the FEMTO-ST Institute in well controlled and understood conditions, fits (2.45) well. The second amplifier was measured at liquid-He temperature at the University of Western Australia [54, Fig. 2]. The original data, given as the equivalent noise temperature as a function of the input power, have been turned into the more readable form of Fig. 2.9. For input powers 0.1–10 µW (−70 to −50 dB m), the $1/f$ phase noise is fairly constant. The broken-line plot (−80 dB m) is clearly polluted by the background noise of the instrument. In the reported experiment the white noise, not shown in Fig. 2.9, was visible only at $f > 10$ kHz.

Mistakes commonly found in amplifier specifications

Numerous data sheets give a specification for white phase noise, in dBc at a suitably large offset frequency, but often without indicating the amplifier's input power. Such

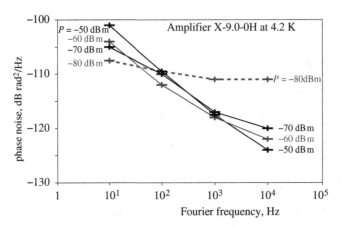

Figure 2.9 Phase noise of the amplifier X-9.0-20H measured at 4.2 K, at different input-power levels. The experimental data are taken from [54, Fig. 2].

a specification is misleading because it hides the nature of the white noise, which is identified better from the noise figure or the noise temperature.

Another common mistake is to specify the flicker noise in terms of the corner frequency f_c. This has been seen in a number of technical articles and data sheets. Once again, the use of f_c is misleading because f_c depends on the power. Instead, phase flickering should be specified in terms of its b_{-1} coefficient.

2.3.4 Environment-originated noise

The phase lag and gain of an amplifier are sensitive to the environment that surrounds it, chiefly the supply voltage and the temperature. Of course, a fluctuating environment generates amplitude noise and phase noise; this mechanism is a parametric modulation. For small fluctuations, the amplifier can be assimilated to a linear time-invariant system (Section 1.5), which is best described by its response $h(t)$ to the Dirac $\delta(t)$ function or equivalently by its frequency response $H(j\omega)$. Letting $T(t)$ be the environment temperature, or any other parameter, the output amplitude and phase are

$$\alpha(t) = T(t) * h_\alpha(t) ,$$ (2.48)

$$\varphi(t) = T(t) * h_\varphi(t) ,$$ (2.49)

where the symbol $*$ is the convolution operator (see subsection 1.5.1). The impulse responses $h_\alpha(t)$ and $h_\varphi(t)$ contain the time lag between environment and effect. Unfortunately, in most practical cases we do not have sufficient information to use (2.48) and (2.49). We must rely on the measured amplitude or phase-noise spectra alone.

Thermal effects

It is observed experimentally that temperature fluctuations give rise also to a spectrum proportional to $1/f^5$, as shown in Fig. 2.10. The frequency scale of this phenomenon depends on the electronics and on the mechanical assembly as well. Though the question is still somewhat unclear, the $1/f^5$ noise seems to be due to the propagation of heat in electronic circuits; this filters the temperature fluctuations of the environment. The $1/f^5$ slope has also been observed in precision dc amplifiers [95, 86]. Better thermal shielding and higher thermal inertia shift the $1/f^5$ region towards lower frequencies. In reality, the $1/f^5$ region is at most a large bump because the low-frequency asymptotic slope cannot be steeper than -1, otherwise the amplifier group delay would diverge rapidly. Another type of bump has been reported [70], associated with positive feedback or "ringing" of the heat propagation inside the amplifier.

2.4 Phase noise in cascaded amplifiers

The phase noise of a cascade of amplifiers is governed by the rules indicated in Fig. 2.11 and detailed below. Although our discussion is in terms of the phase noise, the amplitude noise is governed by the same laws. It turns out that the rules for parametric noise are at first sight counter-intuitive and radically different from any experience one may have had with white noise.

2.4.1 White phase noise

In a chain of amplifiers, the white noise adds up at the input of each device, regardless of the carrier power. Thus, we can add the noise power of each device as referred to the input of the chain and calculate the noise figure F, as in subsection 2.2.2. Then we evaluate the phase noise using $b_0 = F k T_0 / P_0$, (2.30), which holds for additive noise in general regardless of the amplifier's internal structure and number of stages. Therefore, the amplifier is governed by an adaptation of the Friis formula (2.22) to phase noise:

$$(b_0)_{\text{chain}} = \left(F_1 + \frac{F_2 - 1}{A_1^2} + \frac{F_3 - 1}{A_2^2 A_1^2} + \cdots \right) \frac{k T_0}{P_0} \quad \text{(white noise)} . \qquad (2.50)$$

As before, the first stage has the highest weight in the chain.

2.4.2 Flicker phase noise

It was shown in subsection 2.3.2 that, owing to the parametric nature of flicker, the phase-noise spectrum $S_\varphi(f) = b_{-1}/f$ is independent of the carrier power. Thus when m amplifiers are cascaded, each contributes its own phase noise b_{-1}/f. Assuming that the amplifiers are independent, the noise adds up statistically. Hence the noise power is the sum of the individual powers, so that we have

$$(b_{-1})_{\text{chain}} = \sum_{i=1}^{m} (b_{-1})_i \quad \text{(flicker noise)} . \qquad (2.51)$$

It should be noted that *the Friis formula (2.22) is incorrect in this case.* Instead, the phase-noise contribution of each amplifier is determined by the noise parameter b_{-1}, independently of the position of the amplifier in the chain.

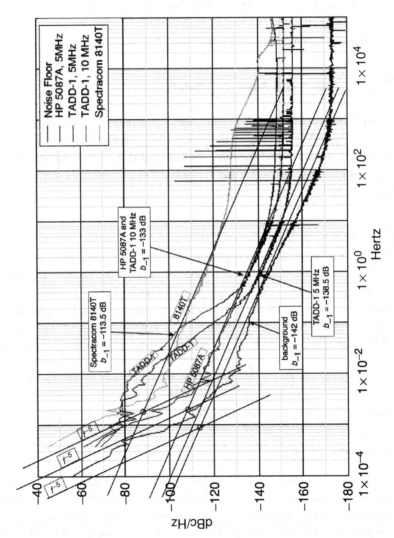

Figure 2.10 Phase noise of some distribution amplifiers, showing the $1/f^5$ noise of environmental origin. The comments are those of the present author. Courtesy of John Ackermann, http://www.febo.com/, 2007.

Additive noise

$$F = F_1 + \frac{F_2 - 1}{|A_1|^2} + \dots$$

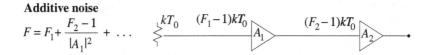

Flicker noise

$$S_\varphi = S_{\varphi 1} + S_{\varphi 2}$$

$$S_\alpha = S_{\alpha 1} + S_{\alpha 2}$$

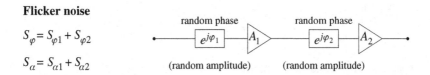

Environmentally originated noise

$$\varphi = \varphi_1 + \varphi_2$$

$$\alpha = \alpha_1 + \alpha_2$$

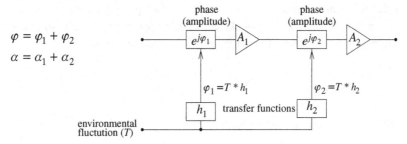

Figure 2.11 Propagation of phase noise in a cascade of amplifiers.

2.4.3 Environment-originated phase noise

According to subsection 2.3.4, this type of noise results from an environmental quantity $T(t)$ that modulates the carrier through an appropriate transfer function $h(t)$. The quantity $T(t)$ can be the temperature, the humidity, the supply voltage, etc. Let us consider a chain of amplifiers located at the same place or in the same box, powered by the same source, etc. All these amplifiers experience the same fluctuating quantity $T(t)$, which gives rise to a phase noise $\varphi(t) = T(t) * h(t)$. Hence, the total phase fluctuation at the output of a chain of m amplifiers is

$$\varphi(t) = T(t) * h_1(t) + T(t) * h_2(t) + \dots + T(t) * h_m(t) \tag{2.52}$$

$$= T(t) * \sum_{i=1}^{m} h_i(t) \tag{2.53}$$

or, equivalently,

$$\Phi(s) = T(s) \sum_{i=1}^{m} H_i(s), \tag{2.54}$$

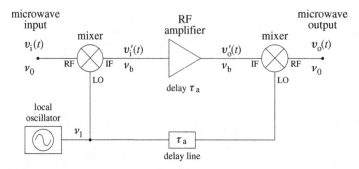

Figure 2.12 Transposed-gain amplifier.

using Laplace transforms.[2] The phase-noise spectrum is obtained as $S_\varphi(f) = |\Phi(f)|^2$, after replacing s by $j2\pi f$, taking the one-sided spectrum, and averaging. The problem with this formal approach is that in virtually all practical cases we have insufficient information. So we can only make calculations for the worst case, in which all phases originating from a single fluctuating parameter add up, so that

$$S_\varphi(f) = \left[\sum_{i=1}^{m} \sqrt{S_{\varphi i}(f)} \right]^2 . \tag{2.55}$$

Otherwise, we have to rely on measurements.

2.5   ★ Low-flicker amplifiers

As a consequence of the Leeson effect, to be introduced in Chapter 3, the $1/f$ phase noise of the sustaining amplifier is of paramount importance for oscillator stability. Therefore, a number of techniques have been developed to improve this parameter.

2.5.1   Transposed-gain amplifiers

The transposed-gain amplifier (Fig. 2.12) originates from the fact that the $1/f$ phase noise of microwave amplifiers is higher than that of microwave mixers and of HF–VHF bipolar amplifiers. To overcome this, the input signal is down-converted from the microwave frequency ν_0 to a suitable frequency $\nu_b = \nu_0 - \nu_l$, amplified, and up-converted back to ν_0. The delay line, which matches the group delay τ_a of the amplifier, is necessary in order to cancel the local oscillator's phase noise, which is common to and coherent at the down-converting and the up-converting mixer. This type of amplifier was proposed as the sustaining amplifier of microwave oscillators independently by Driscoll and Weinert [27] and by Everard and Page-Jones [31].

---

[2] The Laplace transform is denoted by the upper case letter corresponding to the time-domain quantity, except for the pair $T(s) \leftrightarrow T(t)$.

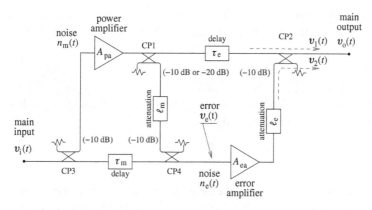

Figure 2.13 Feedforward amplifier, showing the main and error branches.

2.5.2 Feedforward amplifiers

Feedforward and feedback amplifiers were studied around 1930 at the Bell Telephone Laboratories [7, 8] as a means of correcting the distortion in vacuum-tube amplifiers. Feedback has been used extensively since, while feedforward was nearly forgotten. Recently, wireless engineers realized that feedforward can be used to correct the non-linearity of power amplifiers of code-division multiple access (CDMA) networks and other telecommunications systems in which a high peak-to-average power ratio limits the use of highly nonlinear high-efficiency amplifiers.

The scheme of a feedforward amplifier is shown in Fig. 2.13. A general treatment of these amplifiers is available in [76]. Here we focus on the noise properties, which may be derived from three asymptotic cases.

1. In the first case we assume that the whole system is free from noise. Under these conditions, we see that the delay τ_m and the loss ℓ_m (where the subscript refers to the main branch) must be set for the error signal $v_e(t)$ to be equal to zero. For reasons not analyzed here, matching the amplitude and phase of the two paths is not sufficient: it is also necessary to match the group delay [92].
2. In the second case, we add the noise $[n_m(t)]_{pa}$ originating at the input of the power amplifier keeping the error amplifier noise-free. After combining the output of the power amplifier with the main input, the result is an error signal $[n_e(t)]_{pa}$ at the input of the error amplifier. This error signal is amplified and added with an appropriate weight to the delayed output of the power amplifier to form the signal $[v_2(t)]_{pa}$ that nulls the effect of $[n_m(t)]_{pa}$ at the main output. This second case illustrates the noise-cancellation mechanism of the feedforward amplifier.
3. In the third case, we consider a noise-free power amplifier and introduce the noise $[n_e(t)]_{ea}$ originating at the input of the error amplifier. Here, we note that $[n_e(t)]_{ea}$ is amplified and sent to the main output with no compensation mechanism.

The above reasoning should indicate that the noise $[n_m(t)]_{pa}$ originating in the power amplifier is canceled within the accuracy limit of the feedforward path, while the noise

$n_e(t)|_{ea}$ originating in the error amplifier is not canceled. Interestingly, the nature of the noise $n_m(t)|_{ea}$ is still unspecified. This fact has the following consequences.

White noise

Letting $[n_e(t)]_{ea} = F_{ea}kT_0$ be the white noise of the error amplifier, we find that the equivalent noise of the feedforward amplifier is the noise of the error amplifier referred to the main input. Thus, the noise figure is

$$F = F_{ea} \prod_i \ell_i^2 \quad \text{(noise figure)}, \tag{2.56}$$

where $\prod_i \ell_i^2$ is the product of all the power losses, dissipative and intrinsic, in the path from the main input to the input of the error amplifier. For this reason, it is good practice to choose a low coupling factor for CP3 and CP4. For example, a -10 dB coupler has an intrinsic[3] loss of 0.46 dB in the main path. The white phase noise is $b_0 = FkT_0/P_0$, as in any other amplifier.

Distortion

In a well-balanced feedforward amplifier, the distortion of the power amplifier is fully corrected at the coupler CP2. Additionally, there is no carrier power at the input of the error amplifier. Thus, the error amplifier amplifies only the distortion of the power amplifier, for it operates in the low-power regime. Under these conditions the error amplifier is fully linear, so the distortion compensation is nearly perfect. The residual distortion is due to an incorrect feedforward signal $v_2(t)$, i.e. to an incorrect ℓ_e or τ_e, and to a small nonlinearity of the error amplifier due to some residual carrier at its input.

Flicker noise

We have seen that the noise of the power amplifier can be completely removed by the feedforward mechanism. Hence, in this paragraph we analyze only the $1/f$ noise of the error amplifier. Let the signals added in the coupler CP2 be

$$v_1(t) = V_1 \cos 2\pi v_0 t, \tag{2.57}$$

$$v_2(t) = V_2 \cos 2\pi v_0 t + v_x(t) \cos 2\pi v_0 t - v_y(t) \sin 2\pi v_0 t, \tag{2.58}$$

where $v_x(t)$ and $v_y(t)$ are the in-phase and quadrature components of the error-amplifier noise, as in (1.9). For $V_2 \ll V_1$, the amplitude and phase fluctuation at the main output are

$$\alpha(t) = \frac{v_x(t)}{V_1}, \tag{2.59}$$

$$\varphi(t) = \frac{v_y(t)}{V_1}. \tag{2.60}$$

---

[3] The intrinsic loss of a coupler is due to energy conservation.

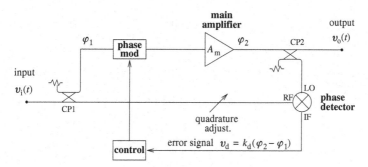

Figure 2.14 Noise-corrected amplifier.

Subsection 2.3.2 shows that, in the case of flickering, the amplitude of the noise sidebands is proportional to the amplitude of the carrier (see (2.34)). Consequently, $v_x(t)$ and $v_y(t)$ can be made arbitrarily small by adjusting ℓ_m and τ_m to minimize V_2. Imperfect setting of ℓ_e or τ_e, as well as some residual carrier at the input of the error amplifier, results in residual flicker.

Oscillator application

If the feedforward amplifier is overdriven, the power amplifier cannot provide the required power. This results in a strong residual carrier at the input of the error amplifier and ultimately in phase and amplitude flickering at the main output.

In oscillator applications, usually the gain control needed for the oscillation amplitude to be constant relies upon saturation in the sustaining amplifier. In this case, the low-flicker feature of the feedforward amplifier, in comparison with traditional amplifiers may be reduced or lost. The obvious solution is to introduce a separate amplitude limiter in the loop, provided that the $1/f$ noise of this limiter is sufficiently small. At microwave frequencies this would seem to be viable using a Schottky-diode limiter because the $1/f$ noise of these diodes is lower than that of microwave amplifiers. In optoelectronic oscillators, it is possible to control the gain by exploiting the nonlinear characteristics of the Mach–Zehnder optical modulator, so that the feedforward amplifier operates in a fully linear regime.

Interestingly, the design of a feedforward amplifier for oscillator applications is simpler than for the general case. The reason is that the signal bandwidth at the amplifier input is narrow, limited by the resonator. Hence, it is sufficient to null the error-amplifier input *at the carrier frequency*, instead of in a large bandwidth. Provided that the group delay is reasonably small, it is therefore sufficient to null the error-amplifier input by adjusting the phase instead of the group delay.

2.5.3 Feedback noise-degeneration amplifiers

The scheme of Fig. 2.14 makes use of a low-noise phase detector that measures the instantaneous phase fluctuation $\varphi_e(t) = \varphi_2(t) - \varphi_1(t)$ of the main amplifier by comparing the amplifier output with the input. The feedback control nulls the error signal

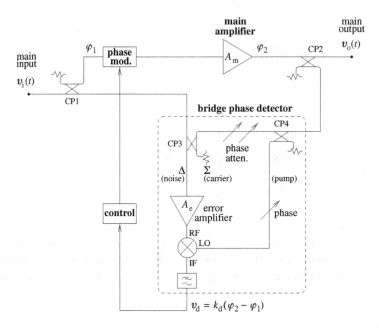

Figure 2.15 Noise-corrected amplifier with bridge phase detector. The phase and amplitude controls needed to stabilize the null at the Δ port are not shown.

and thus compensates for the phase noise of the main amplifier and phase modulator. The control is effective for a bandwidth that can be of order 10–200 kHz. In the control bandwidth, the residual phase noise is chiefly due to the phase detector, including the output preamplifier (not shown). Using a double-balanced mixer as the phase detector, this scheme is effective at microwave frequencies, where the mixer's $1/f$ noise is lower than that of the amplifier. Conversely, improving a HF–VHF amplifier in this way can be more difficult because the $1/f$ noise of amplifiers is already low.

Reference [27] describes a microwave oscillator that makes use of a sustaining amplifier of the type shown in Fig. 2.15, with a double-balanced mixer as the phase detector.

Bridge implementation

The amplifier shown in Fig. 2.15 is a variation on that of Fig. 2.14 in which the double-balanced mixer is replaced by a sophisticated bridge detector. The bridge is balanced by setting the phase and attenuation in the CP4–CP3 path for the carrier to null at the Δ port of CP3. Hence, all the carrier goes into Σ, while only the noise is present in Δ. The noise is amplified and down-converted to dc by synchronous detection. The variable phase at the mixer LO port is set for the detection of phase noise. All details of this low-noise detection method are explained in [82].

In practice, this bridge requires that the phase and attenuation are automatically controlled for the carrier null to be stable in the long run. This control is not shown in Fig. 2.15. When inserting the amplifier into an oscillator, only the amplitude is to be controlled; the phase is set to zero by the oscillation frequency, which follows the drift of the bridge. An oscillator application of this type of amplifier is found in [55].

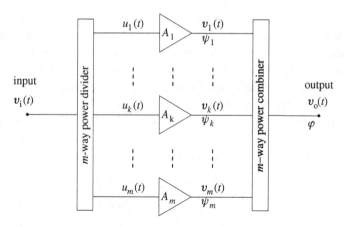

Figure 2.16 Parallel amplifier.

The main virtue of the bridge phase detector is its low $1/f$ noise, which is limited by the variable attenuator and variable phase shifter that balance the bridge. Flicker in the error amplifier is virtually absent because the carrier power at its input is close to zero, and parametric up-conversion of the near-dc flicker (subsection 2.3.2) does not take place. The detector's white noise is $b_0 = F_e k T_0 / P_e$, i.e. the equivalent noise $F_e k T_0$ of the error amplifier referred to the power P_e at the input of the error amplifier. The power loss in the path from the main input to the input of the error amplifier is the product $\prod_i \ell_i^2$ of all the power losses ℓ_i^2, dissipative and intrinsic. Thus, using the power P_0 at the main input, the white phase noise is

$$b_0 = \frac{F_e k T_0}{P_0} \prod_i \ell_i^2 . \tag{2.61}$$

The noise of the corrected amplifier is determined by the phase detector in the bandwidth of the control, and by the noise of the phase modulator and the main amplifier outside this bandwidth. To minimize the flicker, the control bandwidth must be wider than the flicker corner of the main amplifier. The couplers CP1 and CP3 impact on the white noise.

Example 2.3. Let us assume that the noise figure of both amplifiers is 1 dB, that the loss of CP1 and CP3 is 3.5 dB (intrinsic plus dissipative), and that the noise figure of the phase modulator is 1.5 dB, due to a 1.5 dB loss. Under these conditions, the noise figure of the corrected amplifier is 8 dB (due to the error amplifier and two couplers) in the control bandwidth, and 6 dB outside the control bandwidth (one coupler, phase modulator, and the main amplifier).

2.5.4   Parallel amplifiers

A parallel amplifier has m equal amplifiers connected in parallel, as shown in Fig. 2.16. The power splitter and the power combiner are assumed ideal, that is, impedance-matched, symmetrical, and free from dissipative loss. The parallel configuration extends the power output of a given technology because each branch contributes $1/m$ of the output power. Additionally, there are popular configurations that improve the impedance matching by exploiting the properties of the $90°$ power-divider and power combiner. Interestingly, the push–pull amplifier works as a parallel amplifier in which two branches are combined in a $180°$ phase relationship.

Let us first analyze the white phase noise $S_\varphi(f) = b_0$. Denoting the input power by P_0, the power at the input of each branch is P_0/m. Consequently, the phase-noise spectrum is $(b_0)_{\text{branch}} = mFkT_0/P_0$. Combining the output of m independent branches, the phase noise is

$$b_0 = \frac{FkT_0}{P_0} \quad \text{(parallel amplifier)} . \quad (2.62)$$

In summary, the advantage of combining m independent noises is lost because the input power is divided by m and hence the parallel configuration does not reduce the white phase noise. Moreover, actual power dividers show dissipative losses, and this makes the white noise of the parallel amplifier higher than the noise of a branch.

In the case of flicker noise, the coefficient b_{-1} of each branch is not increased by dividing the input power by m. Hence, we expect a noise reduction. The analytical proof is detailed below.

By inspection of Fig. 2.16, recalling that the power splitter and combiner are impedance-matched, symmetrical, and free from dissipative loss, we find that

$$u_k(t) = \frac{1}{\sqrt{m}} v_i(t) \quad \text{(branch-amplifier input)} , \quad (2.63)$$

and

$$v_o(t) = \frac{1}{\sqrt{m}} \sum_{k=1}^{m} v_k(t) \quad \text{(main output)} . \quad (2.64)$$

Letting

$$v_i(t) = V_i e^{j\omega_0 t} \quad \text{(input carrier)} \quad (2.65)$$

be the input carrier, we find $u_k(t) = (1/\sqrt{m})V_i e^{j\omega_0 t}$ at the input of each branch, and $v_k(t) = (a_1/\sqrt{m})V_i e^{j\omega_0 t}$ at the output. The coefficient a_1 is the gain, i.e. the coefficient of the first term in the nonlinear input–output function (2.31). Combining the m carriers, we find

$$v_o(t) = a_1 V_i e^{j\omega_0 t} \quad \text{(output carrier)} \quad (2.66)$$

at the main output.

The proof of the $1/f$ noise reduction is detailed for the nonlinear model of subsection 2.3.2 and briefly summarized for the quasi-linear model. The mathematical manipulations and the results are the same.

Nonlinear model

Introducing the flicker noise, the signal at the output of branch k is

$$v_k(t) = \frac{1}{\sqrt{m}} V_i \left\{ a_1 + 2a_2 \left[n'_k(t) + jn''_k(t) \right] \right\} e^{j\omega_0 t} . \tag{2.67}$$

The above equation is (2.34) adapted to the symbols of Fig. 2.16. The random phase

$$\psi_k(t) = 2\frac{a_2}{a_1} n''_k(t) \tag{2.68}$$

at the output of each branch has a spectrum $S_\psi(f) = [b_{-1}]_{\text{branch}}/f$, with

$$[b_{-1}]_{\text{branch}} = 4\frac{a_2^2}{a_1^2} S_{n''}(f) . \tag{2.69}$$

Each $\psi_k(t)$ contributes to the output phase $\varphi(t)$ by an amount

$$\varphi_k(t) = \frac{1}{m} \frac{V_i \, 2a_2 n''_k(t) \, e^{j\omega_0 t}}{a_1 V_i \, e^{j\omega_0 t}} \tag{2.70}$$

$$= \frac{1}{m} 2\frac{a_2}{a_1} n''_k(t) . \tag{2.71}$$

Thus the contribution of $\varphi_k(t)$ to the output phase spectrum is

$$S_{\varphi, k}(f) = \frac{1}{m^2} 4\frac{a_2^2}{a_1^2} S_{n''_k}(f) . \tag{2.72}$$

Notice that the expression (2.70) for $\varphi_k(t)$ has as its denominator the total output amplitude; thus $\varphi_k(t)$ is the contribution of the branch k to the total output signal. The output phase noise is found by adding the contribution of the m branches:

$$S_\varphi(f) = \sum_{k=1}^{m} \frac{1}{m^2} 4\frac{a_2^2}{a_1^2} S_{n''_k}(f) . \tag{2.73}$$

Therefore we have

$$S_\varphi(f) = \frac{1}{m} 4\frac{a_2^2}{a_1^2} S_{n''}(f) , \tag{2.74}$$

under the assumption that the noise of all branches is equal ($S_{n''_k}(f) = S_{n''}(f)$). Adding the phase-noise spectral densities is a valid mathematical operation because the branches are statistically independent and because the spectral density is an average square quantity. Comparing (2.69) and (2.74), the flicker coefficient of the parallel amplifier must be given by

$$b_{-1} = \frac{1}{m} [b_{-1}]_{\text{branch}} . \tag{2.75}$$

This proves the flicker reduction property of the parallel configuration.

Table 2.2 Specification of low phase-noise amplifiers. The four columns on the right give the phase noise at 10^2, 10^3, 10^4, and 10^5 Hz. From AML Inc. [1]

Amplifier	Parameters				Phase noise vs. f, Hz			
	gain	F	bias	power	10^2	10^3	10^4	10^5
AML812PNA0901	10	6.0	100	9	-145.0	-150.0	-158.0	-159.0
AML812PNB0801	9	6.5	200	11	-147.5	-152.5	-160.5	-161.5
AML812PNC0801	8	6.5	400	13	-150.0	-155.0	-163.0	-164.0
AML812PND0801	8	6.5	800	15	-152.5	-157.5	-165.5	-166.5
unit	dB	dB	mA	dB m	dB rad^2/Hz			

Quasi-linear parametric model
In the case of the quasi-linear parametric model, (2.67) becomes

$$v_k(t) = \frac{1}{\sqrt{m}} V_i A_0 \left[1 + n'(t) + j n''(t) \right] e^{j\omega_0 t}, \qquad (2.76)$$

which is (2.39) adapted to the symbols of Fig. 2.16. Hence it holds that $\psi_k(t) = n_k''(t)$, thus $[b_{-1}]_{\text{branch}} = S_{n''}(f)$. After repeating the same mathematical manipulations as for the nonlinear model, (2.74) becomes

$$S_\varphi(f) = \frac{1}{m} S_{n''}(f), \qquad (2.77)$$

which yields (2.75). This confirms the flicker reduction.

Commercial amplifiers

At the time of writing, AML Communications Inc. [1] has been identified as the sole manufacturer of parallel amplifiers intended for low-phase-noise applications. Actually, the fact that the low phase noise is achieved by paralleling a number of stages is not stated, but the specifications (Table 2.2) leave little doubt that the number of branches connected in parallel scales in powers of 2. Looking at Table 2.2 from top to bottom, at every row the bias current is doubled and the maximum output power increases by 2 dB. That the power increase is of 2 dB instead of 3 dB is sound, if one accounts for the loss and for the asymmetry of the internal passive networks. The gain reduction of about 1 dB at every step reinforces this conclusion.

The phase-noise specifications of the AML amplifiers are somewhat unclear because the $1/f$ and the white region (Fig. 2.17) are not clearly identified. This could be due to a number of reasons, at which we can only guess, such as the use of conservative specifications and production averaging. Nevertheless, some mismatch between the model and the specifications does not change the main fact, that the phase noise decreases by 2.5 dB every time the number of branches is doubled. The missing 0.5 dB may be ascribed to asymmetry in the power splitters and combiners.

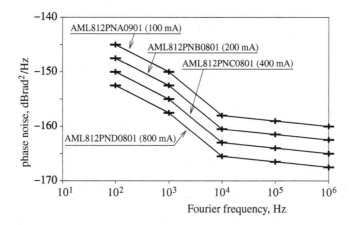

Figure 2.17 Phase noise of a family of AML amplifiers. From AML Communications Inc. [1].

2.5.5   The effect of semiconductor size

Phase flickering is proportional to the inverse physical size of the amplifier's active region. This can be proved through a gedankenexperiment, in which we join the m branches of a parallel amplifier to form a large compound device. An additional hypothesis is required, that the near-dc flickering takes place at the microscopic scale, for there is no correlation[4] between the different regions of the compound device. This hypothesis is consistent with the two most accredited models for flicker noise [49, 69].

2.5.6   SiGe amplifiers

As a matter of fact, well supported by experience, the close-in $1/f$ noise is significantly lower in bipolar transistors than in other types of transistor. This fact is related to their superior linearity and to the greater volume of the region in which amplification takes place. Unfortunately, the gain of Si bipolar transistors drops at microwave frequencies, beyond some 2–3 GHz. This limit has been overcome by recent silicon–germanium technology. A SiGe heterojunction bipolar transistor (SiGe HBT) is similar to a conventional Si bipolar transistor except that the base is made of polycrystalline SiGe, which has a narrower bandgap than Si. For detailed information about SiGe transistors refer to the books [2, 23, 22].

The low $1/f$ noise of SiGe HBTs is of interest in connection with microwave oscillators, where it has been used successfully to implement a variety of high-spectral-purity units spanning from commercial VCOs to research-grade sapphire oscillators [12, 20, 41, 65].

---

[4] Of course, this cannot hold at any scale. We expect that there exists a correlation length characteristic of the lattice or of the defect type.

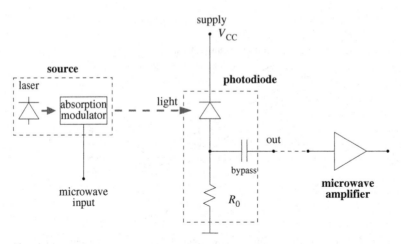

Figure 2.18 Detection of a microwave-modulated laser beam in sinusoidal or distorted regime. Other types of source produce similar intensity modulation.

2.6   ⋆ Detection of microwave-modulated light

In this section we study the noise in the detection of a microwave-modulated laser beam. This topic is of paramount importance for optoelectronic oscillators. We will refer to the sinusoidal or distorted modulation regime (Fig. 2.18). With this restriction, most of the microwave power is in the output bandwidth of the photodiode, so that we can take the fundamental output frequency. It should be pointed out that in the sharp-pulse regime the photodetector operation changes, being dominated by the transient inside the semiconductor. Although it is important in modern optics, this topic is beyond our scope.

The instantaneous power $P_l(t)$ of the optical signal sinusoidally modulated with a modulation index m is

$$P_l(t) = \overline{P}_l(1 + m \cos \omega_0 t) ; \tag{2.78}$$

in this section, we use the subscripts l for "light" and "0" for "microwave." Equation (2.78) is formally similar to the traditional amplitude modulation of radio broadcasting, but here optical power is modulated rather than RF voltage. Thus, if the instantaneous power is described by (2.78) the light's electric field is modulated in both amplitude and phase.

It follows from (2.23) that the detector photocurrent is

$$i(t) = \frac{q\eta}{h\nu_\lambda} \overline{P}_l(1 + m \cos \omega_0 t) . \tag{2.79}$$

Only the ac term $m \cos \omega_0 t$ of (2.79) contributes to the microwave signal. The microwave power fed into the load resistance R_0 is $\overline{P}_0 = R_0 \overline{i_{ac}^2}$, hence

$$\overline{P}_0 = \frac{1}{2} m^2 R_0 \left(\frac{q\eta}{h\nu_l} \right)^2 \overline{P}_l^2 . \tag{2.80}$$

2.6.1 Modulation index

For a given maximum peak power of the laser in Fig. 2.18, the highest microwave power at frequency ω_0 is obtained using square-wave modulation. For our purposes, this condition is equivalent to $m = 4/\pi \simeq 1.273$, which is the first term of the Fourier series of the square wave switching between ± 1.

A case of great practical interest is that of the electro-optic modulator (EOM), the type most used in photonic delay-line oscillators (Chapter 5). The EOM optical transmission, as a function of the driving voltage v, is

$$T = \frac{1}{2} + \frac{1}{2} \sin \frac{\pi v}{V_\pi} , \tag{2.81}$$

V_π being the half-wave voltage of the modulator. Driving the modulator with a microwave signal $v(t) = V_0 \cos \omega_0 t$, the instantaneous transmission is

$$T(t) = \frac{1}{2} \left[1 + 2J_1 \left(\frac{\pi V_0}{V_\pi} \right) \cos \omega_0 t + \cdots \right] , \tag{2.82}$$

where J_1 is the first-order Bessel function of the first kind. Equation (2.82) derives from the $k = 0$ term of the series expansion

$$\sin(z \cos \theta) = 2 \sum_{k=0}^{\infty} (-1)^k J_{2k+1} \cos (2k + 1)\theta . \tag{2.83}$$

The discarded terms "$\cdots$" in (2.82) are harmonics of frequency $n\omega_0$, with integer $n \geq 2$. Though neglected for the output signal at ω_0, their presence is essential in that they ensure that $0 \leq T \leq 1$.

Comparing (2.82) with (2.78), the modulation index is

$$m = 2J_1 \left(\frac{\pi V_0}{V_\pi} \right) . \tag{2.84}$$

The maximum of (2.84) is $m \simeq 1.164$, which occurs at $V_0 = 0.586 V_\pi$.

Harmonic distortion could be avoided by keeping m small. Yet, this is not beneficial because harmonic distortion has no first-order effect on noise and because the microwave harmonics are easily filtered out. However, increasing m increases the microwave power and thus the signal-to-noise ratio. The optical power is limited by saturation in the photodetector. In practice, the microwave power and the dc bias of the EOM are sometimes difficult to set and maintain at the maximum modulation index. This is due to the possibility of bias drift and to the thermal sensitivity of the lithium niobate out of which the EOM is made.

2.6.2 Phase noise

The discrete nature of photons leads to shot noise of power spectral density $S_i(f) = 2q\bar{i}$ (A^2/Hz) at the detector output. It follows from (2.25) that the spectrum of the

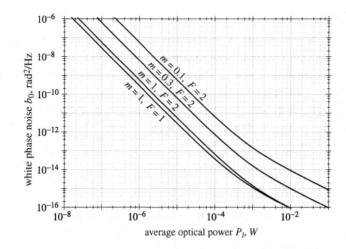

Figure 2.19 White phase noise, as a function of optical power (see (2.87)).

power dissipated by the load resistance R_0 is determined by the dc (average) term of (2.79):

$$N_s = 2R_0 \frac{q^2\eta}{h\nu_l} \overline{P_l} \qquad \text{(shot noise)} . \qquad (2.85)$$

We introduce the subscript "s" for "shot" noise. In virtually all cases the photodetector needs amplification,[5] which introduces the equivalent input noise of the amplifier loaded by R_0 at the temperature T_0, whose power spectral density is

$$N_e = FkT_0 \qquad \text{(amplifier noise, (2.17))} . \qquad (2.86)$$

The total white noise $N_t = N_e + N_s$ leads to a phase-noise floor $S_\varphi(f) = b_0$, with $b_0 = (N_s + N_e)/P_0$. Using (2.80), (2.85), and (2.86), this floor is

$$b_0 = \frac{2}{m^2} \left[\frac{FkT_0}{R_0} \left(\frac{h\nu_l}{q\eta} \right)^2 \frac{1}{\overline{P_l}^2} + 2 \frac{h\nu_l}{\eta} \frac{1}{\overline{P_l}} \right] \qquad \text{(white phase noise)}. \qquad (2.87)$$

Interestingly, the noise floor is proportional to $1/\overline{P_l}^2$ at low power, where the amplifier noise is dominant, and to $1/\overline{P_l}$ at high power. The dominant law changes the threshold power

$$P_{\text{th}} = \frac{1}{2} \frac{FkT_0}{R_0} \frac{h\nu_l}{q^2\eta} \qquad \text{(threshold)} \qquad (2.88)$$

at which the shot noise equals the amplifier equivalent noise. Figure 2.19 shows the noise floor b_0 as a function of the optical power for some reference cases.

---

[5] Actually, a photonic oscillator that does not have explicit microwave amplification has been tested at NIST [71]. The microwave gain mechanism takes place in the photodetector, whose output power is, by virtue of (2.80), proportional to the squared optical power. The absence of a microwave amplifier is accounted for by taking $F = 1$ (0 dB) in our equations.

Example 2.4. At a wavelength 1.55 μm (193 THz), with $\eta = 0.6$, $F = 1$ (noise-free amplifier), and $m = 1$, we obtain a threshold power equal to 689 μW and a noise floor equal to 9.9×10^{-15} rad^2/Hz (-140 dB rad^2/Hz).

Parametric noise is also present; this results from the sum of the photodetector noise and the amplifier noise, with the rules discussed in Section 2.4. In the literature there is very little information about the photodiode $1/f$ phase noise, and no information at all about environmentally originated noise. The flicker noise in microwave photodetectors is expected to be in the region of 10^{-12} rad^2/Hz [93, 88].

Exercises

In these exercises, assume that all circuits are impedance-matched to 50 Ω and that they are in equilibrium at a temperature $T_0 = 290$ K. We make use of the amplifiers shown in the following table.

Amplifier	Gain		Noise F		b_{-1}	
A	8.0	18.0	1.2	0.8	1×10^{-11}	-110
B	4.0	12.0	1.5	1.8	4×10^{-11}	-104
C	10	20.0	3.2	5	1×10^{-10}	-100
D	3.2	10	4	6	1×10^{-12}	-120
unit	V/V	dB	—	dB	rad^2/Hz	dB rad^2/Hz

2.1 Calculate the noise spectrum (in W/Hz) of amplifier A around the arbitrary carrier frequency ν_0, referred to the input and to the output. Calculate the noise temperature referred to the input and to the output. Repeat for amplifiers B–D.

2.2 The power at the input of amplifier A is $P_0 = 20$ μW. Calculate and sketch the phase-noise spectrum $S_\varphi(f)$. Repeat for amplifiers B–D.

2.3 Repeat the previous exercise for input powers 10 μW, 31.6 μW, and 100 μW. For each amplifier, sketch the three spectra on the same plot.

2.4 The amplifiers A and B are now cascaded (A precedes B). Calculate the noise power spectrum (in W/Hz) around the arbitrary carrier frequency ν_0, referred to the input and to the output. Calculate the equivalent noise figure F_a of the chain and the noise temperature. Repeat for amplifiers C and D (C preceding D).

2.5 Repeat the previous exercise with the amplifiers interchanged (now B precedes A and D precedes C).

2.6 The amplifiers A and B are cascaded (A preceding B). Assume 8 mW output power. Calculate and sketch the phase-noise spectrum $S_\varphi(f)$. Repeat for amplifiers C and D (C precedes D).

2.7   A 1 nW carrier is amplified by a cascade of the four different amplifiers A–D. What are the orders of the amplifiers that make the noise figure a minimum or a maximum? For the two cases, calculate the noise figure. Also calculate and sketch the phase-noise spectrum $S_\varphi(f)$.

2.8   Two amplifiers C are connected in parallel with ideal 90° loss-free power dividers and combiners. Calculate and sketch the phase-noise spectrum $S_\varphi(f)$ for input powers 1 mW, 2 mW, and 5 mW. What happens if the loss-free power dividers and combiners are replaced by real ones having dissipative loss 0.5 dB?

2.9   Four amplifiers C are connected in parallel with ideal two-way 90° power dividers and combiners. This means that there are three power dividers at the input and three power combiners at the output. Sketch the scheme. Calculate and sketch the phase-noise spectrum $S_\varphi(f)$ for input powers 1 mW, 2 mW, and 5 mW. What happens if the loss-free power dividers and combiners are replaced by real ones having dissipative loss 0.5 dB?

2.10   A high-speed photodetector of quantum efficiency $\eta = 0.8$ receives a 5 mW light beam at 1.32 μm wavelength. The detector output is connected to a 50 Ω resistor. Calculate the current flowing in the resistor and the shot noise (in W/Hz). Calculate an equivalent noise temperature that accounts for both thermal noise and shot noise.

2.11   In the setup of the previous exercise, the light beam is now intensity modulated by a microwave signal, so that the power swings sinusoidally between 4 mW and 6 mW. Calculate the microwave power in the resistor and the white phase noise.

2.12   As a consequence of temperature fluctuations, the phase-noise spectrum of an amplifier has a term b_{-5}/f^5 with $b_{-5} = 10^{-25}$ rad^2/Hz. What happens if two such amplifiers are cascaded or connected in parallel? And with three such amplifiers cascaded or in parallel?

3 Heuristic approach to the Leeson effect

This chapter introduces the basic noise phenomena in feedback oscillators. The first phenomenon is the Leeson effect, by which the phase noise of the sustaining amplifier – the amplifier that compensates for the resonator insertion loss and ensures stationary oscillation – is turned into frequency noise, causing the phase noise to diverge in the long run. Second, the oscillator is followed by an output buffer, which contributes its own phase noise. Finally, the oscillator tracks the frequency of the internal resonator as well as its frequency fluctuations.

3.1   Oscillator fundamentals

The basic feedback oscillator (Fig. 3.1) is a loop around which the gain A of the sustaining amplifier compensates for the resonator loss at a given frequency frequency ω_0. For the signal to be periodic, i.e. exactly replicated after one trip around the loop, it is necessary that the loop phase arg $A\beta(j\omega)$ be zero or a multiple of 2π at $\omega = \omega_0$. The amplitude and phase conditions for stationary oscillation of a complex signal $A\beta(j\omega)$ are known collectively as the *Barkhausen* condition:

$$|A\beta(j\omega)| = 1 , \tag{3.1}$$

$$\arg A\beta(j\omega) = 0 , \tag{3.2}$$

or equivalently

$$A\beta(j\omega) = 1 \qquad \text{(Barkhausen condition)} . \tag{3.3}$$

The unused input (0 V) on the left-hand side of Fig. 3.1 serves the purpose of mathematical analysis and physical insight. This input is used to set the initial conditions from which oscillation starts and to introduce the equivalent noise of the loop components.

In most practical cases, the resonator frequency response $|\beta(j\omega)|^2$ has a sharp peak at $\omega = \omega_0$. It is therefore convenient to assume that the amplifier gain A is independent of frequency: the small dependence of A on frequency, if any, is moved to $\beta(j\omega)$. The product of functions $A\beta(j\omega)$, which are still unspecified, is shown in Fig. 3.1. For the sake of clarity, we will leave until later the more general case of oscillation at a frequency ω_0 not equal to the resonator's natural frequency ω_n.

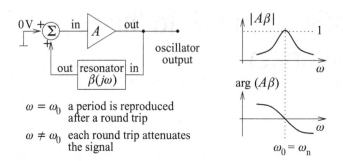

Figure 3.1 Basic feedback oscillator oscillating at the exact natural frequency of the resonator.

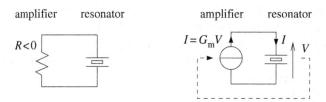

Figure 3.2 Negative-resistance (quartz) oscillator.

The model of Fig. 3.1 is general. It applies to a variety of cases and is not limited to electrical and mechanical systems. A little effort may be necessary to identify A and $\beta(j\omega)$, as follows.

In the case of a two-port microwave cavity connected to an amplifier in a closed loop, matching the real oscillator to the model of Fig. 3.1 is trivial. A less trivial example is the negative-resistance oscillator shown in Fig. 3.2. In this case, the feedback function $\beta(j\omega)$ is the resonator impedance $Z(j\omega) = V(j\omega)/I(j\omega)$. The current $I(j\omega)$ is the input and the voltage $V(j\omega)$ is the output. The resonator impedance is a complex function of frequency that takes a real value (a resistance) at $\omega = \omega_0$. A negative transconductance G_m (here the subscript stands for "mutual") is used to represent the amplifier. We can relate the oscillator of Fig. 3.2 to the general scheme of Fig. 3.1 by observing that the controlled current generator is a transconductance amplifier that senses the voltage V across the resonator and delivers a current $I = G_m V$. Common bipolar transistors and field-effect transistors can be regarded as a current source controlled by the input voltage (Fig. 3.3), and they can be arranged in a scheme that implements a negative conductance.

The generalization of the above concept is that the amplifier input and output signals can be either voltages or currents, which gives four basic feedback topologies, well known in analog electronics [91, Chapter 8].

The question of signs deserves some attention. In circuit theory it is generally agreed that the current is positive when it enters a load or leaves a generator. This goes with the fact that a generator is intended to provide power, and a load to absorb power. Accordingly, when we look at the controlled generator as if it were a resistor, which is a

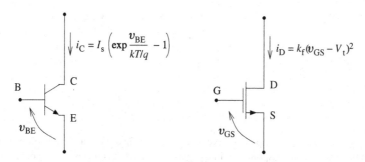

Figure 3.3 Bipolar and field-effect transistors can be regarded as a current source controlled by the input voltage.

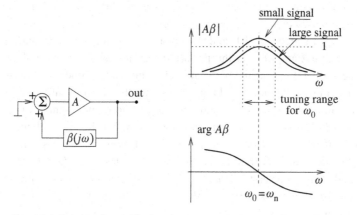

Figure 3.4 Starting the oscillator.

load, we have to change the sign of the current and thus of the resistance. The condition $A\beta(j\omega) = 1$ requires that $G_m > 0$, which is equivalent to a *negative resistance*.

3.1.1 Starting the oscillor

Oscillations start from noise or from a switch-on transient. In the spectrum of such random signals, only a small energy is initially present at ω_0. For the oscillation to grow to a desired amplitude, it is necessary that $|A\beta(j\omega)| > 1$ at $\omega = \omega_0$ for small signals (Fig. 3.4). With $|A\beta(j\omega)| > 1$, the oscillation rises exponentially at a frequency ω_0 defined by arg $A\beta(j\omega) = 0$. As the oscillation amplitude approaches the desired value an amplitude control mechanism (not shown in Fig. 3.4) reduces the loop gain, so that the loop reaches the stationary condition $A\beta(\omega_0) = 1$. Two solutions are preferred in electronic oscillators.

> *AGC.* The amplitude is stabilized by an automatic gain control (AGC) external to the amplifier, which reduces the gain in proportion to the amplitude and with a suitable time constant. This solution is preferred when low harmonic distortion

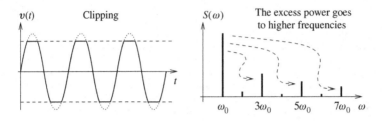

Figure 3.5 When a signal is clipped, the excess power is pushed into harmonics of order higher than 1. In this example the clipping is almost symmetric and thus the even harmonics are strongly attenuated.

is an important feature, i.e. in variable-frequency oscillators where cleaning the output with a filter is impractical. Examples are the bridged-T and the Wien-bridge oscillators used in audio-frequency distorsiometers.

Fast clipping. In precision RF and microwave oscillators, large-signal clipping of the sustaining amplifier is the preferred way to stabilize the amplitude. High-frequency stability is achieved because clipping is a fast process and thus virtually free from phase fluctuations. Figure 3.5 shows the effect of saturation. When the input amplitude exceeds the saturation level, the output signal is clipped. The actual behavior can be a smooth compression rather than hard clipping. Further increasing the input level, the gain at ω_0 decreases and the excess power is pushed into the harmonics, at frequencies that are multiples of ω_0. Quartz oscillators, in which the amplitude is stabilized by amplifier saturation, can attain the remarkable stability of a few parts in 10^{-7}, with a flicker spectrum in the region of 1 Hz [79].

In summary, it is important to understand that, for real-world oscillators:

1. It is necessary that $|A\beta(j\omega)| > 1$ for small signals, otherwise the oscillation will not start.
2. The condition $|A\beta(j\omega)| = 1$ results from large-signal gain saturation. This is necessary for the oscillation amplitude to be stable and allows the amplitude of the oscillation to reach a steady state.
3. The oscillation frequency is determined only by the phase condition $\arg A\beta(j\omega) = 0$. This is necessary for the oscillation to be regenerated after a round trip.

3.1.2   Pulling the oscillation frequency

To explain how an oscillator can be tuned to the desired frequency, we need the conceptual distinction between the oscillation frequency ω_0 and the natural frequency ω_n of the resonator. Of course, the former generally lies close to the latter. The mathematical details are left until Chapter 4.

A simple way to adjust the oscillation frequency is to introduce into the loop a static phase ψ, external to the resonator (Fig. 3.6). This method is often preferred at microwave frequencies, where line stretchers are available.

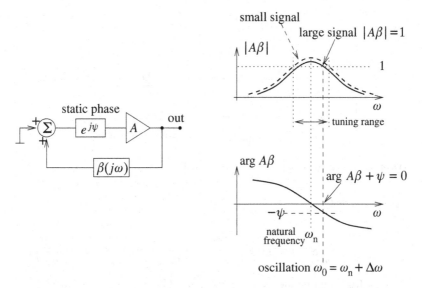

Figure 3.6 Tuning the oscillation frequency by the insertion of a static phase.

After introducing ψ, the Barkhausen phase condition becomes

$$\arg A\beta(j\omega) + \psi = 0 \qquad \text{at } \omega = \omega_0 . \tag{3.4}$$

Hence, the loop oscillates at the frequency

$$\omega_0 = \omega_n + \Delta\omega \tag{3.5}$$

for which

$$\arg A\beta(j\omega) = -\psi . \tag{3.6}$$

For reference, $\psi > 0$ means that the loop leads; thus the oscillator is pulled to a frequency higher than the exact resonance ($\Delta\omega > 0$).

Of course, it is necessary that the oscillator saturates. Accordingly, the maximum tuning range is the frequency range in which

$$|A\beta(j\omega)| > 1 \qquad \text{(small signal)} \tag{3.7}$$

is guaranteed, so that the gain can be reduced by saturation. Out of this range, the energy stored in the loop decays exponentially and hence no oscillation is possible.

In the linear region (see the lower right-hand diagram in Fig. 3.6), it holds that

$$\Delta\omega = -\frac{\psi}{d[\arg A\beta(j\omega)]/d\omega} . \tag{3.8}$$

If the resonator is a simple circuit governed by a second-order differential equation with low damping factor (i.e. a large quality factor Q), in the vicinity of the natural frequency

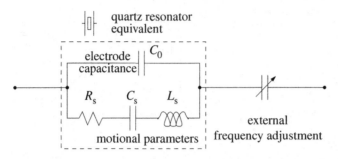

Figure 3.7 Typical tuning scheme for quartz oscillators.

ω_n it holds that

$$\frac{d}{d\omega} \arg A\beta(j\omega) = -\frac{2Q}{\omega_n} \quad \left(\text{or } \frac{d}{d\nu} \arg A\beta(\nu) = -\frac{2Q}{\nu_n} \right); \qquad (3.9)$$

thus the fractional frequency offset introduced by the static phase ψ is

$$\frac{\Delta\omega}{\omega_0} = \frac{\Delta\nu}{\nu_0} = \frac{\psi}{2Q} \quad \text{for} \quad \frac{\Delta\omega}{\omega_0} \ll \frac{1}{2Q}. \qquad (3.10)$$

Alternative tuning method

A second tuning method consists of pulling the natural frequency of the resonator by modifying the parameters of the resonator's differential equation. The adjustment circuit is no longer distinct from the resonator. In this case, there is no need to introduce the static phase ψ. This method is often used in quartz oscillators, where a variable capacitor is used to alter the resonator's natural frequency (Fig. 3.7).

3.2   The Leeson formula

Let us consider a loop oscillator in which the feedback circuit $\beta(j\omega)$ is an ideal resonator with a large[1] quality factor Q, free from frequency fluctuations (Fig. 3.8). The relaxation time of the resonator is

$$\tau = \frac{Q}{\pi} T_0 = \frac{Q}{\pi \nu_0} = \frac{2Q}{\omega_0}. \qquad (3.11)$$

Replacing the static phase ψ of (3.8) by a time-varying phase $\psi(t)$ results in an oscillator output phase that is a function of time:

$$v(t) = V_0 \cos[\omega_0 t + \varphi(t)]. \qquad (3.12)$$

In the case of *slow fluctuations* of $\psi(t)$, slower than the inverse of the relaxation time τ, the phase $\psi(t)$ can be treated as a quasi-static perturbation. Equation (3.10) tells us

---

[1] Strictly speaking, only $Q \gtrsim 10$ is necessary.

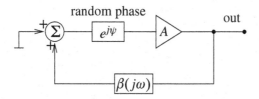

Figure 3.8 The phase noise of the amplifier and of all other components of the loop is modeled as a random phase ψ at the input of the amplifier.

that the oscillator responds to it with a frequency fluctuation

$$(\Delta v)(t) = \frac{v_0}{2Q} \psi(t) , \tag{3.13}$$

with associated power spectral density

$$S_{\Delta v}(f) = \left(\frac{v_0}{2Q} \right)^2 S_\psi(f) . \tag{3.14}$$

The instantaneous output phase is

$$\varphi(t) = 2\pi \int (\Delta v)(t) \, dt . \tag{3.15}$$

The time-domain integration corresponds to a multiplication by $1/j\omega = 1/(j2\pi f)$ in the Fourier transform, thus into a multiplication by $1/(2\pi f)^2$ in the spectrum. The factor 2π in (3.15) cancels with the 2π in $1/(j2\pi f)$. Consequently, the oscillator's slow phase-fluctuation spectrum is

$$S_\varphi(f) = \frac{1}{f^2} \left(\frac{v_0}{2Q} \right)^2 S_\psi(f) . \tag{3.16}$$

For the *fast components* of $\psi(t)$, i.e. those faster than the inverse of the relaxation time τ, the resonator is a flywheel that steers the signal. Loosely speaking, the resonator does not respond to fast phase fluctuations; its output signal is a pure sinusoid. Accordingly, the fluctuation $\psi(t)$ goes through the amplifier and shows up at its output without being fed back to the input. No noise regeneration takes place in this conditions, thus

$$\varphi(t) = \psi(t) \tag{3.17}$$

and

$$S_\varphi(f) = S_\psi(f) . \tag{3.18}$$

By adding the effect of fast and slow fluctuations ((3.16) and (3.18)), we get the Leeson formula relating the oscillator's output phase spectrum to the amplifier's phase

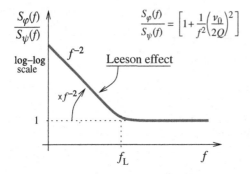

Figure 3.9 The modulus squared of the Leeson-effect phase transfer function $H(jf)$.

fluctuations:

$$S_\varphi(f) = \left[1 + \frac{1}{f^2}\left(\frac{v_0}{2Q}\right)^2\right] S_\psi(f) \qquad \text{(Leeson formula)}. \qquad (3.19)$$

The Leeson formula (3.19) can be rewritten as

$$S_\varphi(f) = \left(1 + \frac{f_L^2}{f^2}\right) S_\psi(f), \qquad (3.20)$$

where

$$f_L = \frac{v_0}{2Q} = \frac{1}{2\pi\tau} \qquad \text{(Leeson frequency)} \qquad (3.21)$$

is the Leeson frequency. By inspection of (3.19), it can be seen that the oscillator behaves as a *first-order filter* with a perfect integrator (a pole in the origin in the Laplace transform representation) and a cutoff frequency f_L (a zero on the real left-hand axis in the complex plane). The theory of linear[2] systems describes an oscillator in terms of its phase-noise transfer function $H(jf)$; we have

$$|H(jf)|^2 = \frac{S_\varphi(f)}{S_\psi(f)} = 1 + \frac{1}{f^2}\left(\frac{v_0}{2Q}\right)^2 \qquad \text{(Fig. 3.9)}. \qquad (3.22)$$

Some final remarks deserve attention. We have obtained the Leeson formula by following a heuristic approach based on physical insight. The rigorous proof, given in Chapter 4, confirms this result. Equation (3.19), and equivalently (3.20), is a mathematical property of the oscillator loop and is independent of the physical origin of the phase fluctuations $\psi(t)$, which are normally addressed as phase noise. The origin of phase noise is still unspecified, and the noise of the resonator has still not been accounted for.

---

[2] Real oscillators are inherently nonlinear. Nonetheless in most practical cases phase noise is a small perturbation, and a linear analysis of the amplitude and phase is correct.

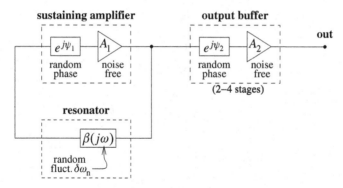

Figure 3.10 Block diagram of a real oscillator consisting of resonator, sustaining amplifier, and buffer.

The formula (3.19) was proposed by David B. Leeson [63] as a model for short-term frequency fluctuations and was initially intended to explain the phase noise of the crystal-oscillator frequency-multiplier source for airborne Doppler radar applications. As will be shown in Chapter 6, for the range of the baseband frequency f used in radar systems [64], it is appropriate to ascribe essentially all the noise to the amplifier.

3.3   The phase-noise spectrum of real oscillators

The Leeson effect has been introduced in terms of a transfer function governing the propagation of the amplifier phase noise to the phase noise at the oscillator output. Now we will analyze the phase noise of a real oscillator consisting of the three main blocks shown in Fig. 3.10, the *resonator*, the *sustaining amplifier*, and the *output buffer*. The last is necessary for output isolation and, ultimately, to prevent the external load affecting the oscillation frequency. Each block contributes its own noise to the oscillator output. Here, it is sufficient to account for the amplifier's white and flicker phase noise, because the environmentally originated noise shows up only at very low Fourier frequencies (subsection 2.3.4 and Fig. 2.10), where resonator instability turns out to be the dominant phenomenon.

3.3.1   Noisy amplifier and fluctuation-free resonator

For a given noisy amplifier, the phase noise is white at higher frequencies and of the flicker type below the corner frequency f_c. When such an amplifier is inserted into an oscillator, the result for the oscillator signal, before buffering, is the two basic spectra seen in Fig. 3.11.

Type-1 spectrum $(f_L > f_c)$

This is the most frequently encountered spectrum, often found in microwave oscillators and in high-frequency piezoelectric oscillators ($\gtrsim 100$ MHz), in which f_L is made

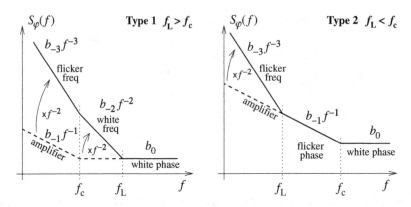

Figure 3.11 With a flickering amplifier, the Leeson effect yields two types of spectrum. A noise-free resonator and buffer are provisionally assumed. Adapted from [83] and used with the permission of the IEEE, 2007.

high by the high oscillation frequency and the low quality factor Q. By inspecting Fig. 3.11 (left) from the right-hand side to the left-hand side, we first encounter the white phase noise $b_0 f^0$ of the amplifier transferred to the oscillator output and then the white frequency noise $b_{-2} f^{-2}$ due to the Leeson effect on the amplifier white noise. At lower frequencies, where the amplifier flickering shows up, the oscillator noise turns into the frequency-flicker type $b_{-3} f^{-3}$. The phase flickering $(b_{-1} f^{-1})$ is not observed at the output.

Type-2 spectrum ($f_L < f_c$)

This spectrum is found in RF (2.5–10 MHz) high-stability quartz oscillators and in cryogenic microwave oscillators. In both cases, the resonator exhibits a high quality factor. Looking at Fig. 3.11 (right) again from the right-hand side to the left-hand side, the amplifier's phase noise changes from white $(b_0 f^0)$ to flicker $(b_{-1} f^{-1})$ at $f = f_c$. Phase flickering $(b_{-1} f^{-1})$ shows up in the frequency region from f_L to f_c. The Leeson effect occurs only below f_L, where the oscillator noise turns into frequency flickering $(b_{-3} f^{-3})$. The white frequency noise $(b_{-2} f^{-2})$ is not observed at the output.

   Comparing the two spectra, there can be either the f^{-1} or the f^{-2} noise type. The presence of *both*, sometimes observed, can be explained using a more sophisticated model which accounts for the output buffer's noise contribution.

3.3.2   The effect of the output buffer

Adding a buffer to the oscillator loop, we notice that the buffer and oscillator are independent. Still neglecting the effect of the environment, the buffer's phase-noise spectrum contains only white and flicker noise while the oscillator's spectrum becomes significantly higher at $f < f_L$ because of the Leeson effect. Thus, assuming that the buffer and the sustaining amplifier make use of similar technology, the buffer noise will show up only at $f \gtrsim f_L$. In the $f > f_L$ region, there is no phase feedback in the

Table 3.1 The effect of the buffer on the output phase-noise spectrum if all amplifiers employ the same technology

No. of buffer stages	Oscillator flicker	
1	$(b_{-1})_{osc} = 2(b_{-1})_{ampli}$	3.0 dB
2	$(b_{-1})_{osc} = 3(b_{-1})_{ampli}$	4.8 dB
3	$(b_{-1})_{osc} = 4(b_{-1})_{ampli}$	6.0 dB
4	$(b_{-1})_{osc} = 5(b_{-1})_{ampli}$	7.0 dB
5	$(b_{-1})_{osc} = 6(b_{-1})_{ampli}$	7.8 dB

loop, and the phase noise at the output of the loop is the phase noise of the sustaining amplifier (see (3.17)). Consequently the sustaining amplifier and the buffer are modeled as cascaded amplifiers, and the associated noise spectra are added according to the rules given in Section 2.4. To account for buffer noise, the noise spectra of Fig. 3.11 are replaced by those of Figure 3.12, as explained below.

White phase noise

In the white-noise region, the phase-noise spectrum at the loop output is $b_0 = FkT_0/P_0$, (2.30). The buffer phase noise is governed by the same law but with a higher input power in the denominator because of the sustaining amplifier. Therefore, the sustaining amplifier has the highest weight in summing the phase noise spectra, as deduced from (2.50). Under these conditions, with careful design the phase-noise contribution of the buffer can be made negligible.

Flicker phase noise

In the case of phase flickering, the amplifier noise is roughly independent of the carrier power, hence (2.50) does not apply. Instead, the noise spectrum is the sum of the single spectra, as given by (2.51). While the sustaining amplifier often consists of a single stage, two or more buffer stages may be needed for isolation. Because of the Leeson effect, if a sophisticated low-flicker amplifier (Section 2.5) is affordable then it should be used as the sustaining amplifier, not as the buffer. These design considerations yield the conclusion that the $1/f$ noise of the output buffer is expected to be higher than that of the sustaining amplifier.

If similar technology is employed, the flicker coefficient b_{-1} is about the same for each stage, whether it occurs in the sustaining amplifier or in the buffer. Hence, if there are n buffer stages plus one sustaining amplifier, the total $1/f$ noise spectrum observed at the output is $n + 1$ times the $1/f$ noise spectrum of one amplifier (Table 3.1). In the absence of any other information, we can guess that an oscillator will have one sustaining amplifier and three buffer stages, for the estimated flicker of the sustaining amplifier is

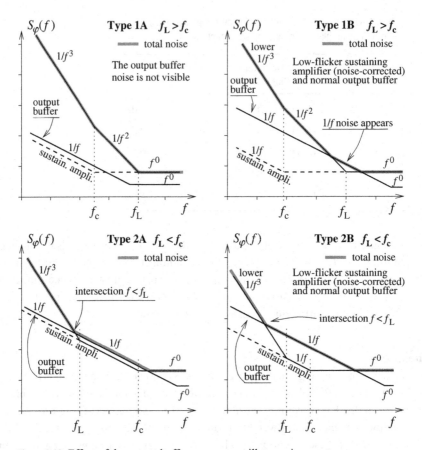

Figure 3.12 Effect of the output buffer on some oscillator noise spectra.

one-quarter (-6 dB) of the total noise at the oscillator output, with an error of 1 dB if there are actually two or four buffers.

Type-1A spectrum ($f_L > f_c$)

The sustaining amplifier and the buffer make use of the same technology, for they have similar flicker characteristics. Yet the Leeson effect, occurring at $f_L > f_c$, turns the white noise of the sustaining amplifier into $1/f^2$ noise for $f < f_L$. It is seen in Fig. 3.12 that the Leeson effect makes the buffer flickering negligible. Hence, the insertion of a buffer at the output of the loop leaves the spectrum unaffected.

The type-1A spectrum is typical of everyday microwave oscillators such as dielectric resonance oscillators (DROs) and the yttrium iron garnet (YIG) based oscillators.

Type-1B spectrum ($f_L > f_c$)

For maximum stability, the oscillator loop employs a noise-degeneration amplifier (sub-section 2.5.3), which exhibits reduced flicker. Conversely, the buffer is a traditional amplifier because the cost and complexity of the noise-degeneration amplifier are justified only inside the loop, where the Leeson effect takes place. As a consequence, the larger $1/f$ noise of the buffer is expected to show up. Interestingly, this is the only type of spectrum that contains *both* $1/f$ and $1/f^2$ noise types.

The larger buffer $1/f$ noise hides the f^0 to $1/f^2$ transition, which is the signature of the Leeson effect. Of course, the Leeson frequency can still be estimated by extrapolating the f^{-2} segment. The continuation of the f^{-2} line crosses the white phase noise at $f = f_L$.

The type-1B spectrum is observed in some exotic low-noise microwave oscillators that make use of a cryogenic resonator, for maximizing Q, and a noise-degeneration amplifier.

Type-2A spectrum $(f_L < f_c)$

The region between f_L and f_c is affected by buffer flickering, which is higher than the $1/f$ noise of the sustaining amplifier because of the higher number of stages. As a consequence, the corner where the f^{-1} noise turns into f^{-3} noise is pushed to a frequency lower than the true f_L. This corner frequency has the same graphical signature as the Leeson effect and is easily misinterpreted.

The type-2A spectrum can be found in high-stability 5–10 MHz quartz oscillators, and in microwave cryogenic oscillators. It is the only spectrum in which there is a clearly visible $1/f$ region of the spectrum and in which similar technology is employed for sustaining the amplifier and buffer. In this case, we can infer the $1/f$ noise of the sustaining amplifier by subtracting 6 dB from the $1/f$ output phase noise, under the assumption that there are three buffer stages (Table 3.1).

Type-2B spectrum $(f_L < f_c)$

The type-2B spectrum differs from type 2A in that the noise-degeneration amplifier makes the $1/f^3$ noise lower. Yet the graphical pattern is the same. The spectrum does not contain $1/f^2$ noise. The $1/f$ noise is the phase noise of the output buffer, which hides the Leeson effect.

One may expect to find the type-2B spectrum in some sophisticated 5–10 MHz quartz oscillators, where f_L is of order 1–10 Hz. However, noise-degeneration amplifiers are not employed in this type of oscillator because the resonator noise is higher than the $1/f^3$ noise originating from the Leeson effect. This will be analyzed thoroughly in Chapter 6.

3.3.3 The effect of resonator noise

Thermal noise is inherent in the dissipative loss of the resonator. This effect is included in the equivalent noise at the input of the amplifier, as explained in Section 2.2. Other noise phenomena, always present and far more relevant to the oscillator's stability, are the flickering and the random walk of the resonant frequency. The fractional-frequency spectral density $S_y(f)$ shows a term $h_{-1}f^{-1}$ for the frequency flicker and a term $h_{-2}f^{-2}$

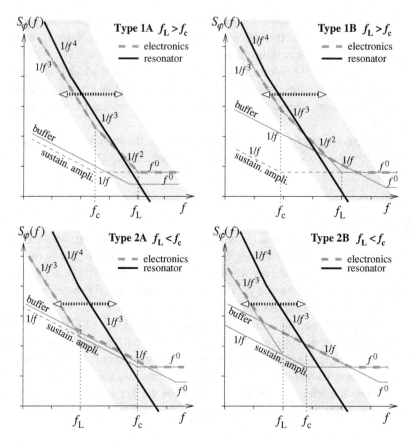

Figure 3.13 Effect of resonator frequency fluctuations on the oscillator noise.

for the frequency random walk; $S_\varphi(f)$ and $S_y(f)$ are related by (1.72):

$$S_\varphi(f) = \frac{v_0^2}{f^2} S_y(f). \tag{3.23}$$

Thus the term $h_{-1} f^{-1}$ of the resonator frequency fluctuation yields a term $b_{-3} f^{-3}$ in the phase noise, while the term $h_{-2} f^{-2}$ of the former yields a term $b_{-4} f^{-4}$ in the latter. Of course, the resonator noise is independent of anything else in the oscillator and thus adds to the noise of the electronics, which includes the buffer and sustaining amplifier.

After introducing the resonator noise, the basic spectra of Figure 3.12 turn into those of Fig. 3.13. The purpose of this figure is to show the graphical signature of all possible cases. For this reason the resonator noise, shown as a bold solid line of slope $1/f^3$ and $1/f^4$, is somewhat arbitrary. As indicated, it can lie anywhere within the light grey regions. The main feature of Fig. 3.13 is the presence of the resonator $1/f^3$ noise, which can be higher or lower than the $1/f^3$ noise due to the Leeson effect. This identifies the resonator, or the sustaining amplifier noise through the Leeson effect, as the main cause of frequency flickering.

In type-1 spectra, the resonator $1/f^3$ noise may hide f_c, but not f_L, and may cross the $1/f^2$ noise due to the Leeson effect (thick grey broken line). In this case, the spectrum

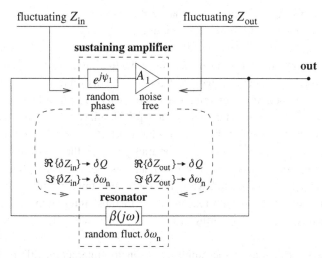

Figure 3.14 Effect of the impedance fluctuation of the sustaining amplifier.

shows a corner point (f^{-3} to f^{-2}) that is easily mistaken for f_c because it shows the same graphical signature. Conversely, in the case of type-2 spectra the resonator $1/f^3$ noise may hide f_L and cross the $1/f$ noise of the buffer and of the sustaining amplifier (thin grey broken line). The spectrum has the same graphical signature as the Leeson effect, i.e. the slope changes from f^{-3} to f^{-1} at a corner point. Of course, this corner frequency is *not* the Leeson frequency. This behavior is typical of high-Q HF quartz oscillators (5–10 MHz). If the resonator $1/f^3$ noise is higher than shown, it may hide *both* f_c and f_L. In this case, only one corner point is visible on the plot, where the resonator noise (f^{-3}) crosses the white noise of the amplifier. This behavior is typical of VHF quartz oscillators.

In all cases, the $1/f^4$ noise (due to frequency random walk) of the resonator is the dominant process at sufficiently low frequencies. At even lower frequencies other processes show up, such as frequency drift and aging.

3.3.4 ⋆ Resonator–amplifier interaction

In the previous sections, the sustaining amplifier has been implicitly considered as a perfectly directional device that transfers the input signal to the output, at most introducing a randomly fluctuating phase in the transfer but having no effect on the resonator parameters. However, that things may not be this simple is suggested by the observation that coupling the resonator to the oscillator inherently introduces dissipation. As a consequence, the resonator's quality factor is lowered by a factor of approximately 0.8–0.3. This is observed in a variety of resonators that include the quartz resonator, the microwave cavity, either dielectric-loaded or not, the whispering-gallery optical resonator, etc. Of course, if the real part of the amplifier's input or output impedance fluctuates then the quality factor also fluctuates. This is shown in Fig. 3.14. Similarly, if the imaginary part of the amplifier's input or output impedance fluctuates then the resonator's natural frequency also fluctuates, and so does the oscillation frequency.

For example, in a 5 MHz quartz oscillator, the pulling capacitance can affect the fractional frequency by $10^{-7}/\text{pF}$. This means that a fluctuation of 5×10^{-19} F would account for a fractional-frequency fluctuation of 5×10^{-14}. This value constitutes a record stability for a quartz oscillator. For reference, in fundamental capacitance-metrology, a resolution of 10^{-10} in the measurement of a 1 pF Thompson–Lampard standard is definitely not out of reach, which means a resolution of 10^{-22} F. The high resolution of the metrology methods probably cannot be transposed to the measurement of the sustaining amplifier, for two reasons. The first reason is that the Thompson–Lampard standard is a four-terminal capacitor, while the sustaining-amplifier input or output is a two-terminal network. The second reason is that the sustaining amplifier should be measured in actual conditions, oscillating in a closed loop, which is clearly a problematic requirement.

The noise mechanism due to the amplifier–resonator interaction differs from the amplifier's input–output phase noise in that the amplifier's fluctuating impedance enters in the resonator dynamic parameters instead of in the feedback, thus it has no cutoff at the offset frequency ω_L. However, this noise mechanism is not reported in the literature. Therefore proving or disproving its relevance in precision oscillators is an open research problem.

3.4   Other types of oscillator

3.4.1   Delay-line oscillator

A delay line (Fig. 3.15) in the feedback path can be the element that determines the oscillation frequency, rather than a resonator. In the frequency domain, a delay τ is described by $\beta(j\omega) = e^{-j\omega\tau}$. Thus, the loop can sustain the oscillation at any frequency ω_l for which $\arg A\beta(j\omega) = 0$, that is,

$$\omega_l = \frac{2\pi l}{\tau} \quad \text{or} \quad \nu_l = \frac{l}{\tau}, \quad \text{for integer } l. \tag{3.24}$$

A selector circuit, not shown in Fig. 3.15, is therefore necessary for the selection of a specific oscillation frequency ω_0.

In quasi-static conditions, the Leeson effect can still be derived from (3.8), repeated here:

$$\Delta\omega = -\frac{\psi}{d[\arg A\beta(j\omega)]/d\omega}. \tag{3.25}$$

For a delay τ, it holds that $d[\arg A\beta(j\omega)] = -\tau$. To this extent, the delay line is equivalent to a resonator of resonant frequency ν_l and quality factor

$$Q = \pi\nu_l\tau; \tag{3.26}$$

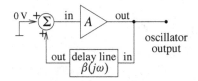

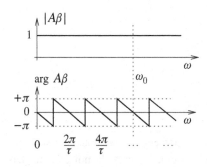

Feedback sustains oscillation
at any ω at which $\arg A\beta = 0$

A selector circuit (not shown)
is needed to select the
oscillation frequency

Figure 3.15 Basic delay-line oscillator.

thus

$$f_{\mathrm{L}} = \frac{1}{2\pi\tau} . \tag{3.27}$$

For slow phase fluctuations it holds that

$$(\Delta\nu)(t) = \frac{\psi(t)}{2\pi\tau} \qquad \text{for } f \ll f_{\mathrm{L}} , \tag{3.28}$$

and therefore

$$S_\varphi(f) = \frac{1}{f^2} \frac{1}{4\pi^2\tau^2} S_\psi(f) \qquad \text{for } f \ll f_{\mathrm{L}} . \tag{3.29}$$

Conversely, this oscillator's response to fast frequency fluctuations is far more complex than in the case of a resonator. In fact, the delay line is a wide-band device and thus it does not stop the fast phase fluctuations. Chapter 5 is devoted to this topic.

3.4.2   Frequency-locked oscillators

This type of oscillator (Fig. 3.16) consists of a voltage-controlled oscillator (VCO) frequency-locked to a passive frequency reference, usually a resonator. In this subsection, we denote by ω_0 the oscillator frequency and by ω_n the resonator's natural frequency. The error signal ν_e is proportional to the frequency error $\Delta\omega = \omega_0 - \omega_n$ of the VCO, not to the phase error. The reason is that the resonator's transfer function $\beta(j\omega)$ turns the frequency fluctuations into phase fluctuations. Therefore, the control is a *frequency-locked loop* (FLL), not to be mistaken for a phase-locked loop (PLL). The FLL, less well known than the PLL, is commonly used in atomic frequency standards and in lasers.

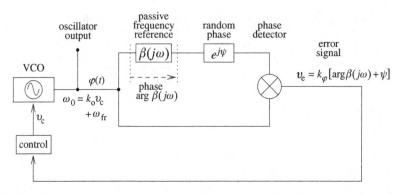

Figure 3.16 Discriminator-stabilized oscillator.

The random phase $\psi(t)$ of Fig. 3.16 represents the residual noise of the system, originating in the lowest-power parts of the circuit. It can be the noise of the phase detector or of the preamplifier (not shown) that follows the detector. We analyze this scheme in quasi-static conditions, thus assuming that the random fluctuation $\psi(t)$ is slower than the relaxation time of the frequency reference. The error signal is $v_e = k_\varphi[\arg \beta(j\omega) + \psi]$, where k_φ is the gain of the phase detector. Using a resonator as the frequency reference, close to the resonance it holds that $\arg \beta = -2Q(\Delta\omega/\omega_n)$. Thus, the error signal is

$$v_e(t) = k_\varphi\left[-2Q\frac{(\Delta\omega)(t)}{\omega_n} + \psi(t)\right]. \tag{3.30}$$

The VCO is governed by the law $\omega_0 = k_o v_c + \omega_{fr}$, where k_o is the gain in rad/(V s), v_c is the control voltage, and ω_{fr} is the free-running frequency at $v_c = 0$. A dc voltage added to v_c, which may be needed to center the system variables in their dynamic range when the error signal is zero, is ignored here because it has no impact on the noise. Denoting by k_c the transfer function of the control, still unspecified, and introducing the VCO law (see below (3.30)) and the loop gain $k_L = k_c k_o k_\varphi$ into (3.30), we find

$$(\Delta\omega)(t) = k_L\left[-2Q\frac{(\Delta\omega)(t)}{\omega_n} + \psi(t)\right]; \tag{3.31}$$

thus

$$(\Delta\omega)(t) = \frac{k_L}{1 + k_L(2Q/\omega_n)}\psi(t). \tag{3.32}$$

Substituting $\Delta\omega = 2\pi\,\Delta\nu$ and $\omega_n = 2\pi\nu_n$ and assuming that $k_L(2Q/\nu_n) \gg 1$ (large control gain), the VCO error (3.32) turns into

$$(\Delta\nu)(t) = \frac{\nu_n}{2Q}\psi(t). \tag{3.33}$$

This result is the same as (3.13), derived for a feedback oscillator in quasi-static conditions. Hence

$$S_\varphi(f) = \frac{1}{f^2}\left(\frac{\nu_0}{2Q}\right)^2 S_\psi(f). \tag{3.34}$$

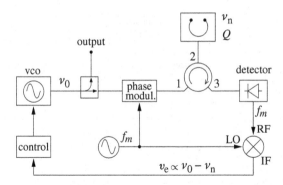

Figure 3.17 Pound oscillator [77]. Adapted from [85] and used with the permission of the IEEE, 2007.

Once again we find the Leeson effect, i.e. the frequency-to-phase conversion of noise that shows up as a multiplication by $1/f^2$ in the left-hand region of the phase-noise spectrum. The complete Leeson formula (3.19), derived for the feedback oscillator, contains the white phase-noise term "1," dominant beyond f_L, which is not present here. Something similar also happens in the case of the frequency-stabilized oscillator. Yet the present case is more complex, because the gain of the phase detector can drop beyond f_L and because the VCO has its own white noise. Hence the structure of the frequency-stabilized oscillator must be detailed for a complete evaluation of its phase-noise spectrum to be possible.

3.4.3 Pound stabilized oscillators

A popular implementation of the discriminator-stabilized oscillator is the Pound scheme [77], shown in Fig. 3.17. The modulation frequency f_m is significantly higher than the resonator bandwidth v_n/Q, thus the modulation sidebands $v_0 \pm f_m$ are completely reflected. The carrier v_0 is partially reflected. The imaginary part of the reflected carrier is proportional to the frequency error $v_0 - v_e$, with a nearly linear law in a small interval. Combining the sidebands and carrier in the power detector, which has a quadratic response, an error signal of frequency f_m is present at the detector output. This signal, downconverted to dc by synchronous detection, controls the VCO to null the error $v_0 - v_e$ in the closed loop. The real part of the reflected carrier, governed by the impedance mismatch at the resonator input, yields a dc signal, which is not detected.

The main point of Pound stabilization is that the use of phase modulation moves the frequency-error information from near-dc to the modulation frequency f_m, which is *far from the flicker region of the electronics*. This feature reduces the Leeson effect dramatically.

Another advantage of the Pound scheme is that the lengths of the microwave cables (from the VCO to the circulator, and from the circulator to the resonator) cancel in the frequency-stabilization equations. As a relevant consequence, the *length fluctuations* impact only on the carrier phase, not on the frequency. In cryogenic oscillators, this fact

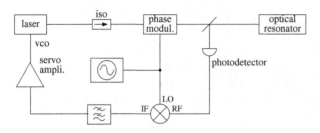

Figure 3.18 Pound–Drever–Hall laser frequency control [26].

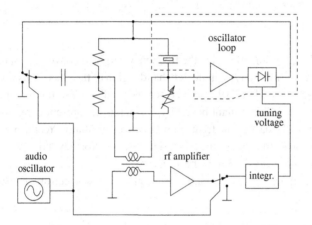

Figure 3.19 Sulzer oscillator [98].

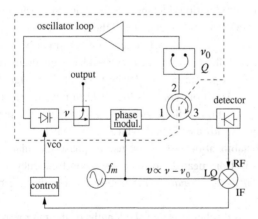

Figure 3.20 Pound–Galani oscillator [34]. Adapted from [85] and used with the permission of the IEEE, 2007.

makes it possible to use room-temperature electronics, far from the cooled resonator, at a reasonably small loss of stability.

A version of the Pound scheme adapted to the stabilization of a laser frequency [26, 6], is shown in Fig. 3.18.

The frequency stabilization inherently requires a VCO locked to the passive reference. It turns out that the stability of this VCO is a critical issue in high-demanding applications. A smart solution is the Sulzer oscillator [98], shown in Fig. 3.19. In this scheme, the same resonator is used as the resonator of the VCO and as the passive frequency reference to which the VCO is locked.

The Sulzer oscillator was invented to solve the problem of high flicker in the early transistors used in quartz oscillators. In fact, the frequency-stabilization feedback loop compensates for the phase flickering of the sustaining amplifier and thus reduces the Leeson effect. Of course, the frequency-error detection works at the audio-frequency f_m, far from the flicker region of the electronics. However smart, Sulzer stabilization is no longer used because the phase flickering of modern transistors is low enough to keep the Leeson effect below the frequency fluctuations of the quartz resonator. This will be analyzed thoroughly in Chapter 6.

The microwave version of the Sulzer oscillator, known as the Pound–Galani oscillator [34], is shown in Fig. 3.20. Of course, at microwave frequencies the Leeson effect is still a major factor limiting oscillator stability. The Pound–Galani is one of the preferred schemes for cryogenic oscillators because the VCO benefits from the high Q of the cryogenic resonator.

4 Phase noise and feedback theory

The main purpose of this chapter is to prove and generalize the Leeson formula (3.19), which was obtained with heuristic reasoning in Chapter 3. This extension in our knowledge suggests new simulation and experimental techniques and enables the analysis of other cases of interest not considered in the current literature, such as mode degeneracy or quasi-degeneracy in resonators or in an oscillator pulled off the resonant frequency. The analysis of delay-line oscillators and lasers in Chapter 5 is based on the ideas introduced here.

Before tackling this proof, however, we must build up a set of tools to manipulate the oscillator phase noise using Laplace transforms and the general formalism of linear time-invariant systems. The underlying idea is to represent the oscillator as a noise-free system that accepts a phase noise $\Psi(s)$ at the input and delivers a phase noise $\Phi(s)$ at the output, as shown in Fig. 4.1. In this way the oscillator may be described by its phase-noise transfer function. The input noise, of course, is the noise of the oscillator's internal parts. The use of a Laplace transform to analyze the phase fluctuation of a sinusoidal signal is inspired by the field of phase-locked loops (PLLs), where it is a common way of calculating the transient response. However, this powerful approach constitutes a new departure in the noise analysis of oscillators. Consequently, a little patience is required as we go through a few sections of mathematics, which at the end will be gathered into the Leeson formula.

4.1 Resonator differential equation

4.1.1 Homogeneous equation

A large variety of resonant systems are described by a second-order homogeneous differential equation of the form

$$\frac{d^2}{dt^2} i(t) + \frac{\omega_n}{Q} \frac{d}{dt} i(t) + \omega_n^2 i(t) = 0 \qquad \text{(resonator)}, \qquad (4.1)$$

where $i(t)$ is the current (or another variable that describes the specific system), ω_n is the natural frequency, and Q is the quality factor. The conditions $\omega_n > 0$ and $Q > 0$ are necessary for the resonator to be physically realizable. These conditions relate to

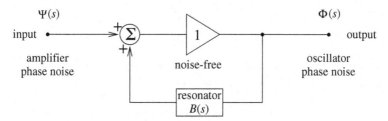

Figure 4.1 Input–output phase-noise model of an oscillator.

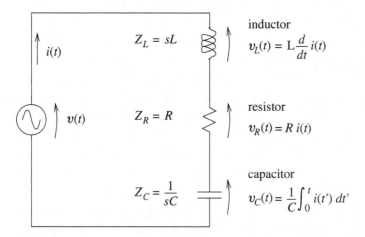

Figure 4.2 An RLC series resonator.

the fact that (4.1) originates from a positive-coefficient system, which in the case of the electrical resonator means $L > 0$, $C > 0$, and $R > 0$, and to the physical need for the resonator's internal energy to decay.

The symbols are inspired by the RLC series resonator, which is used as a canonical example in this chapter. In this case, (4.1) follows immediately from Fig. 4.2 on removing the generator, i.e. on setting $v_L(t) + v_R(t) + v_C(t) = 0$, with the replacements

$$\omega_n^2 = \frac{1}{LC} \qquad \text{(natural frequency } \omega_n) \tag{4.2}$$

and

$$Q = \frac{\omega_n L}{R} = \frac{1}{\omega_n RC} \qquad \text{(quality factor } Q) . \tag{4.3}$$

Other more complex resonant systems can be approximated by (4.1) in the region around ω_n. Distributed systems, such as the microwave cavity resonator and the Fabry–Pérot etalon, are also locally approximated by (4.1).

We will search for the solutions of (4.1) in the time domain. The importance of this approach is that no more than a minimum of knowledge about differential equations is needed. We will guess a solution of the form[1]

$$i(t) = I_0 e^{st} \quad \text{with} \quad s = \sigma + j\omega , \tag{4.4}$$

because the complex exponential e^{st} is an eigenfunction of the derivative operator, that is, $(d/dt) e^{st} = s e^{st}$. After substituting $i(t) = I_0 e^{st}$ in (4.1),

$$s^2 I_0 e^{st} + \frac{\omega_n}{Q} s I_0 e^{st} + \omega_n^2 I_0 e^{st} = 0 , \tag{4.5}$$

we drop the time-dependent term $I_0 e^{st}$. This is possible because (4.5) holds for any time t. In this way, we obtain the associated algebraic equation

$$s^2 + \frac{\omega_n}{Q} s + \omega_n^2 = 0 , \tag{4.6}$$

whose solutions are

$$s = -\frac{\omega_n}{2Q} \pm \sqrt{\Delta} = -\frac{\omega_n}{2Q} \pm \sqrt{\frac{\omega_n^2}{4Q^2} - \omega_n^2} . \tag{4.7}$$

The discriminant Δ, defined as

$$\Delta = \frac{\omega_n^2}{4Q^2} - \omega_n^2 , \tag{4.8}$$

nulls for $Q = \pm\frac{1}{2}$. For $|Q| < \frac{1}{2}$ it holds that $\Delta > 0$, hence the solutions of (4.6) are real.

The case $Q < 0$ is of no physical interest because it describes a resonator whose internal energy builds rather than decaying.

For $0 < Q < \frac{1}{2}$, the solutions of (4.6) are real and negative. The solutions are real because the discriminant is positive. They are negative by virtue of Descartes' rule of signs,[2] after observing that all the coefficients of (4.6) are positive.

For $Q > \frac{1}{2}$, which is always true for resonators of practical interest, it holds that $\Delta < 0$. By virtue of Descartes' rule of signs, the solutions of (4.6) are complex conjugate with a negative real part. After some rearrangement, the solutions of (4.6) are written as

$$s_p, s_p^* = -\frac{\omega_n}{2Q} \pm j\omega_n \sqrt{1 - \frac{1}{4Q^2}} , \qquad Q > 1/2 . \tag{4.9}$$

A relevant property of (4.6) is that the complex conjugate solutions are on a circle of radius ω_n centered at the origin. This is easily seen by applying Pythagoras'

---

[1] In this section we denote the complex variable by $s = \sigma + j\omega$ because of the formal similarity to the results obtained with the Laplace transform. In spite of this similarity, the variable s does not need to be identified with the Laplace complex frequency.

[2] A real-coefficient polynomial of degree m has m roots which are real or occur in complex conjugate pairs. After arranging the polynomial in descending order of the variable and negating the coefficients of odd-power terms, Descartes' rule of signs states that the number of negative roots is the same as the number of sign changes in the coefficients, or a multiple of 2 less than this. In the case of complex conjugate roots, for a pair of sign changes in the coefficients the polynomial has a pair of complex conjugate roots whose real part is negative.

theorem to s_p:

$$R^2 = [\Re\{s_p\}]^2 + [\Im\{s_p\}]^2 \tag{4.10}$$

$$= \frac{\omega_n^2}{4Q^2} + \omega_n^2 \left(1 - \frac{1}{4Q^2}\right) \tag{4.11}$$

$$= \omega_n^2 . \tag{4.12}$$

The same holds for $s_p{}^*$ because $[\Re\{s\}]^2 = [\Re\{s^*\}]^2$ and $[\Im\{s\}]^2 = [\Im\{s^*\}]^2$.

Now we turn our attention back to the homogeneous equation (4.1). Since the coefficients are real, the solutions must be real functions of time. Thus, using $\cos x = \frac{1}{2}(e^{jx} + e^{-jx})$ and $\sin x = -\frac{1}{2}j(e^{jx} - e^{-jx})$, the solutions of (4.1) take the form of a damped oscillation:

$$i(t) = \mathscr{A} \cos \omega_p t \, e^{-t/\tau} + \mathscr{B} \sin \omega_p t \, e^{-t/\tau} \qquad \text{(solution of (4.1))} \tag{4.13}$$

where $\mathscr{A}$ and $\mathscr{B}$ are real constants determined by the initial conditions, and

$$\tau = \frac{2Q}{\omega_n} \qquad \text{(relaxation time)}, \tag{4.14}$$

$$\omega_p = \omega_n \sqrt{1 - \frac{1}{4Q^2}} \qquad \text{(free-decay pseudofrequency)} . \tag{4.15}$$

For high-quality-factor resonators, it holds that

$$\omega_p \simeq \omega_n \left(1 - \frac{1}{8Q^2}\right) \qquad \text{for } Q \gg 1 \tag{4.16}$$

and so

$$\frac{\omega_p - \omega_n}{\omega_n} \simeq -\frac{1}{8Q^2} \qquad \text{for } Q \gg 1 . \tag{4.17}$$

Equation (4.13) requires that $\tau > 0$ for the resonator energy to decay, which explains why the condition $Q > 0$ is necessary. The relaxation time τ is related to the other resonator parameters by the following useful formulae:

$$\tau = \frac{Q}{\pi} T_n = \frac{Q}{\pi \nu_n} = \frac{2Q}{\omega_n} = \frac{1}{\omega_L} = \frac{1}{2\pi f_L} \qquad \text{(relaxation time)}, \tag{4.18}$$

where f_L is the Leeson frequency. An example of damped resonator oscillations is shown in Fig. 4.3.

4.1.2 Inhomogeneous equation

Adding a forcing term $v(t)$ to the series resonator of Fig. 4.2, the resonator is described by the inhomogeneous equation

$$\frac{d^2}{dt^2} i(t) + \frac{\omega_n}{Q} \frac{d}{dt} i(t) + \omega_n^2 i(t) = \frac{1}{L} \frac{d}{dt} v(t) . \tag{4.19}$$

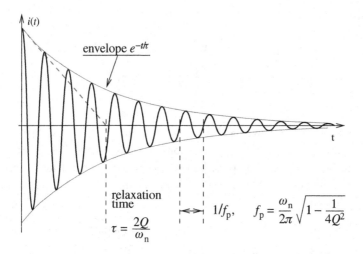

Figure 4.3 Resonator damped oscillation.

We restrict our attention to the case of sinusoidal forcing of frequency ω_0. Of course, the asymptotic response is a sinusoid of the same frequency ω_0 and of appropriate amplitude and phase. This is necessarily so because for $t \to \infty$ the energy contained in the initial conditions is dissipated exponentially with a time constant equal to half the relaxation time. Owing to linearity, the forcing response is added to the homogeneous solution. Hence the complete solution of (4.19) is of the form

$$i(t) = \mathscr{A} \cos \omega_p t \, e^{-t/\tau} + \mathscr{B} \sin \omega_p t \, e^{-t/\tau} + \mathscr{C} \cos \omega_0 t + \mathscr{D} \sin \omega_0 t \,, \qquad (4.20)$$

where $\mathscr{A}$, $\mathscr{B}$, $\mathscr{C}$, and $\mathscr{D}$ are real constants and ω_p is the free-decay angular pseudo-frequency. For a given resonator, $\mathscr{A}$ and $\mathscr{B}$ are the same coefficients as those of the solution (4.13) of the homogeneous equation, determined by the initial conditions. However, the coefficients $\mathscr{C}$ and $\mathscr{D}$ are determined by the forcing term.

4.2   Resonator Laplace transform

In linear circuit theory, the use of the Laplace transform is common because it gives access to a simple and powerful formalism for manipulating the network functions, i.e. the admittances and transfer functions. Using the Laplace transform, an RLC network is described by a *rational function* of the complex variable $s = \sigma + j\omega$ that is *completely determined by its roots* (poles and zeros) on the complex plane. More precisely, a function $F(s)$ having N zeros $s_z = \sigma_z + j\omega_z$ and M poles $s_p = \sigma_p + j\omega_p$ can be written as

$$F(s) = C \, \frac{\prod(s - s_z)}{\prod(s - s_p)} \,, \qquad (4.21)$$

where the product in the numerator is taken over the N zeros and the product in the denominator is over the M poles, and where C is a constant that is determined by the residues. It may be remarked that the terms $s - s_z$ and $s - s_p$ are interpreted as the distances of the point s from the zeros s_z and from the poles s_p. This makes the evaluation of $|F(j\omega)|^2$ a simple application of Pythagoras' theorem:

$$|F(j\omega)|^2 = C^2 \, \frac{\prod \left[\sigma_z^2 + (\omega - \omega_z)^2 \right]}{\prod \left[\sigma_p^2 + (\omega - \omega_p)^2 \right]} \, . \tag{4.22}$$

Using the RLC series resonator as an example, the admittance is

$$Y(s) = \frac{1}{L} \, \frac{s}{s^2 + \omega_n s / Q + \omega_n^2} \, , \tag{4.23}$$

with

$$\omega_n^2 = \frac{1}{LC} \quad \text{and} \quad Q = \frac{\omega_n L}{R} = \frac{1}{\omega_n RC} \, . \tag{4.24}$$

This follows immediately from $Z(s) = sL + R + 1/sC$, as seen in Fig. 4.2. In the sinusoidal regime, for $Q \gg \frac{1}{2}$ the admittance has its maximum $|Y(s)| = Q/\omega L = 1/R$ at $s = j\omega_n$.

For the sake of generalization, we prefer a transfer function of the form

$$\beta(s) = \frac{\omega_n}{Q} \, \frac{s}{s^2 + \omega_n s / Q + \omega_n^2} \quad \text{(general resonator transfer function)} \, , \tag{4.25}$$

which is obtained from (4.23) after multiplication by $\omega_n L / Q$. The function $\beta(s)$ is interpreted as a dimensionless *transfer function* normalized for $|\beta(j\omega_n)| = 1$, as in Chapter 3. With this choice, the Barkhausen condition for stationary oscillation is met with an amplifier of gain $A = 1$. Of course, an impedance or an admittance can still be regarded as a transfer function, with the current as the input and the voltage as the output or vice versa. A number of other resonant systems of interest are described, or well approximated around the frequency ω_n, by a function like (4.25). Figure 4.4 shows the resonator transfer function in the frequency domain and the roots in the complex plane.

Owing to the formal similarity of the denominator of $\beta(s)$ to (4.6), the poles of $\beta(s)$ have the same properties as the solutions of (4.6), namely

1. The poles are real and negative for $0 < Q \le \frac{1}{2}$.
2. The poles are complex conjugates for $Q > \frac{1}{2}$.
3. For $Q > \frac{1}{2}$, the poles are on a circle of radius ω_n centered at the origin.

When the poles are complex conjugates, the system function $\beta(s)$ is often rewritten as

$$\beta(s) = \frac{\omega_n}{Q} \, \frac{s}{(s - s_p)(s - s_p^*)} \quad \text{(resonator poles } s_p, s_p^* = \sigma_p \pm j\omega_p) \tag{4.26}$$

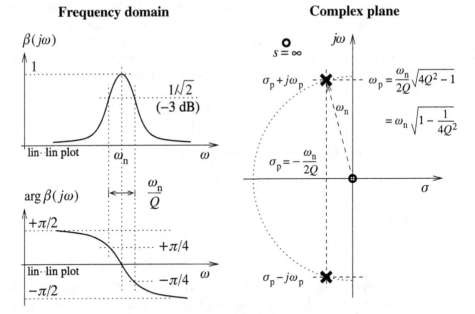

Figure 4.4 Resonator transfer function $\beta(s)$, (4.25), for $Q \gg 1/2$. As usual in complex analysis, the zeros are represented by circles and the poles by crosses on the complex plane.

with

$$\sigma_p = -\frac{\omega_n}{2Q}, \tag{4.27}$$

$$\omega_p = \frac{\omega_n}{2Q}\sqrt{4Q^2 - 1} \tag{4.28}$$

$$= \omega_n\sqrt{1 - \frac{1}{4Q^2}}. \tag{4.29}$$

The frequency response is found by substituting $s = j\omega$ in $\beta(s)$:

$$\beta(j\omega) = \frac{\omega_n}{Q} \frac{j\omega}{-\omega^2 - j2\sigma_p\omega + \sigma_p^2 + \omega_p^2}. \tag{4.30}$$

Defining the *dissonance* χ as

$$\chi = \frac{\omega}{\omega_n} - \frac{\omega_n}{\omega} \qquad \text{(definition of the dissonance } \chi\text{)}, \tag{4.31}$$

we find the relationships shown in Table 4.1. In the vicinity of the natural frequency, the following approximation holds:

$$\chi \simeq 2\frac{\omega - \omega_n}{\omega_n} \qquad \text{for} \qquad \left|\frac{\omega - \omega_n}{\omega_n}\right| \ll 1. \tag{4.32}$$

Unfortunately, this approximation can be used only at positive frequencies. However, the negative-frequency properties can be obtained by symmetry.

Table 4.1 Relevant resonance parameters

$\chi = \dfrac{\omega}{\omega_n} - \dfrac{\omega_n}{\omega}$	dissonance
$\beta(j\omega) = \dfrac{1}{1 + jQ\chi}$	frequency response
$\Re\{\beta(j\omega)\} = \dfrac{1}{1 + Q^2\chi^2}$	real part
$\Im\{\beta(j\omega)\} = -\dfrac{Q\chi}{1 + Q^2\chi^2}$	imaginary part
$|\beta(j\omega)| = \dfrac{1}{\sqrt{1 + Q^2\chi^2}}$	modulus
$\arg\beta(j\omega) = -\arctan Q\chi$	argument

4.2.1 Symmetry properties of high-Q resonators

The function $\beta(s)$, as any network function, has the following properties:

$$\beta(s) = \beta^*(s^*),\tag{4.33}$$

$$|\beta(j\omega)| = |\beta(-j\omega)| \qquad \text{(the modulus is even)},\tag{4.34}$$

$$\arg\beta(j\omega) = -\arg\beta(-j\omega) \qquad \text{(the argument or phase is odd)},\tag{4.35}$$

$$\Re\{\beta(j\omega)\} = \Re\{\beta(-j\omega)\} \qquad \text{(the real part is even)},\tag{4.36}$$

$$\Im\{\beta(j\omega)\} = -\Im\{\beta(-j\omega)\} \qquad \text{(the imaginary part is odd)}.\tag{4.37}$$

These conditions are necessary for $\beta(s)$ to be a real-coefficient analytic function, or by extension a function whose series expansion has real coefficients of s, and ultimately for the inverse transform (i.e. the time-domain impulse response) to be a real function of time. Figure 4.5 shows the symmetry in the case of a very high quality factor, $Q \ggg \frac{1}{2}$. The most remarkable fact is that the resonance shape is almost symmetrical. Thus the symmetry properties (4.34)–(4.37) are also a local approximation around the resonant frequency ω_n and, of course, around $-\omega_n$. Introducing the frequency offset δ, it holds that

$$\begin{aligned}|\beta(j(-\omega_n - \delta))| &\simeq |\beta(j(-\omega_n + \delta))| \\ |\beta(j(\omega_n - \delta))| &\simeq |\beta(j(\omega_n + \delta))|\end{aligned} \qquad \text{(the modulus is even)},\tag{4.38}$$

$$\begin{aligned}\arg\beta(j(-\omega_n - \delta)) &\simeq -\arg\beta(j(-\omega_n + \delta)) \\ \arg\beta(j(\omega_n - \delta)) &\simeq -\arg\beta(j(\omega_n + \delta))\end{aligned} \qquad \text{(the phase is odd)},\tag{4.39}$$

$$\begin{aligned}\Re\{\beta(j(-\omega_n - \delta))\} &\simeq \Re\{\beta(j(-\omega_n + \delta))\} \\ \Re\{\beta(j(\omega_n - \delta))\} &\simeq \Re\{\beta(j(\omega_n + \delta))\}\end{aligned} \qquad \text{(the real part is even)},\tag{4.40}$$

$$\begin{aligned}\Im\{\beta(j(-\omega_n - \delta))\} &\simeq -\Im\{\beta(j(-\omega_n + \delta))\} \\ \Im\{\beta(j(\omega_n - \delta))\} &\simeq -\Im\{\beta(j(\omega_n + \delta))\}\end{aligned} \qquad \text{(the imaginary part is odd)}.\tag{4.41}$$

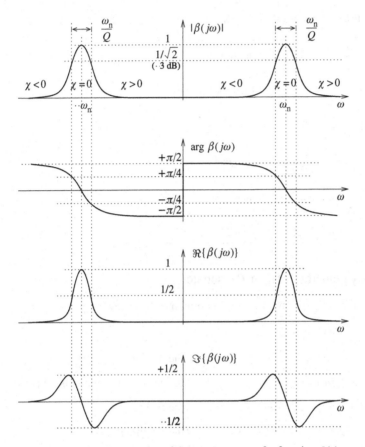

Figure 4.5 Symmetry properties of the resonator transfer function $\beta(s)$.

These properties are a consequence of the proximity of the poles to the imaginary axis. In this case, by virtue of (4.21), when the frequency axis is swept around the resonant frequency ω_n only the distance from the pole close to $j\omega_n$ changes abruptly and symmetrically; the other roots of $\beta(j\omega)$ are not much affected because they are far away. The same thing happens when the frequency axis is swept around $-\omega_n$. By the way, (4.21) also explains why arg $\beta(j\omega)$ flips sign at $\omega = 0$, where $\beta(j\omega)$ has a zero.

4.3   The oscillator

Figure 4.6(a) shows an oscillator loop represented using a Laplace transform. The amplifier of gain A (constant) and the feedback path $\beta(s)$ are blocks with which we are familiar after Chapter 3. The signal $V_i(s)$ at the input of the summing block Σ allows initial conditions and noise to be introduced into the loop. Interestingly, it is also possible to introduce a sinusoid of frequency close to the free-running frequency; this describes an injection-locked oscillator.

(a) Oscillator

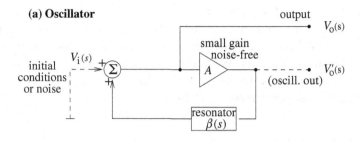

(b) Classical control

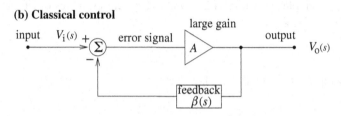

Figure 4.6 Similarity and difference between an oscillator and a classical control.

The topology of the oscillator loop shown in Fig. 4.6(a) is similar to that of the basic scheme used in classical control theory (Fig. 4.6(b)). The analogy enables a straightforward derivation of the oscillator equation from control theory. From this standpoint, an oscillator can be regarded as an ill-designed simple control, made to oscillate by (intentionally) inappropriate feedback. The main difference between the oscillator and the control is that the oscillator has unity-gain positive feedback instead of large-gain negative feedback; this is emphasized by the sign "+" at the feedback input of Σ. In this figure the oscillator output V_o is taken at the amplifier input instead of at the amplifier output. The purpose of this seemingly weird choice is to simplify the equations. Of course, in actual implementations the oscillator output must be the amplifier output V_o' in order to minimize the perturbation to the loop. Additionally, a real resonator has an insertion loss, which is compensated by taking an amplifier gain larger than unity.

Referring to Fig. 4.6(a), elementary feedback theory tells us that the oscillator transfer function, defined as

$$H(s) = \frac{V_o(s)}{V_i(s)} \qquad \text{(definition of } H(s)\text{)}, \tag{4.42}$$

is

$$H(s) = \frac{1}{1 - A\beta(s)} \qquad \text{(see Fig. 4.6(a))}. \tag{4.43}$$

If the denominator $1 - A\beta(s)$ nulls for $s = \pm j\omega_0$, the system provides a finite response for a zero input, i.e. a stationary oscillation of frequency ω_0. From this standpoint, one may regard the oscillator as a system with a pair of imaginary conjugate poles excited by suitable initial conditions and interpret the frequency stability as the stability of the poles on the complex plane. Of course, the condition $1 - A\beta(s) = 0$ for $s = \pm j\omega_0$ is the Barkhausen condition introduced in Chapter 3.

4.3.1   Mathematical properties

We first study the properties of $H(s)$, (4.43), for the case when $\beta(s)$ is the transfer function of a simple resonator, (4.25), and the system oscillates exactly at the natural frequency of the resonator:

$$\omega_0 = \omega_n . \tag{4.44}$$

Thus, substituting (4.25) into (4.43), we get

$$H(s) = \frac{s^2 + (\omega_n/Q)s + \omega_n^2}{s^2 + (1 - A)(\omega_n/Q)s + \omega_n^2} , \tag{4.45}$$

which can be rewritten as

$$H(s) = \frac{s^2 - 2\sigma_p s + \omega_n^2}{s^2 + 2\sigma_p(A - 1)s + \omega_n^2} \tag{4.46}$$

because $\sigma_p = -\omega_n/(2Q)$, (4.27). The above $H(s)$ is a rational function with real coefficients, for it can be written as $\mathcal{N}(s)/\mathcal{D}(s)$, i.e. numerator/denominator where $\mathcal{N}(s)$ and $\mathcal{D}(s)$ are second-degree polynomials. The function $H(s)$ has two poles, either real or complex conjugates depending on the gain A. The root locus is shown in Fig. 4.7. The poles have the following properties.

1. The poles are complex conjugates for $1 - 2Q < A < 1 + 2Q$ and are real elsewhere.
2. When the poles are complex conjugates, they lie on a circle of radius ω_n centered at the origin.
3. The poles are imaginary conjugates for $A = 1$.

The proof is given below.

The poles of $H(s)$ are the zeros of its denominator, i.e. the solutions of $\mathcal{D}(s) = 0$:

$$s_1, s_2 = -\sigma_p(A - 1) \pm \sqrt{\Delta\mathcal{D}} \tag{4.47}$$

$$= -\sigma_p(A - 1) \pm \sqrt{\sigma_p^2(A - 1)^2 - \omega_n^2} . \tag{4.48}$$

The poles are real or complex conjugates depending on the sign of the denominator discriminant $\Delta\mathcal{D}$, defined as

$$\Delta\mathcal{D} = \sigma_p^2(A - 1)^2 - \omega_n^2 . \tag{4.49}$$

(a) Oscillator transfer function $H(s)$ **(b) Detail of the denominator of $H(s)$**

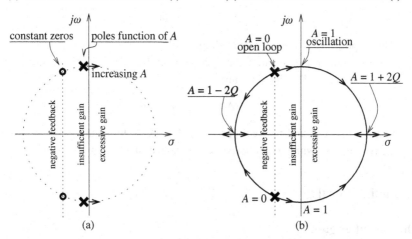

Figure 4.7 (a) Noise transfer function $H(s)$, (4.45), and (b) root locus of the denominator of $H(s)$ as a function of the gain A.

We first solve $\Delta \mathcal{D} = 0$, which yields

$$A_1,\ A_2 = 1 \pm \frac{\omega_n}{\sigma_p} \tag{4.50}$$

and hence

$$A_1,\ A_2 = 1 \mp 2Q \tag{4.51}$$

because $\sigma_p = -\omega_n/(2Q)$, (4.27). As the coefficient of A^2 in $\Delta \mathcal{D}$ is positive, we conclude that

$$\Delta \mathcal{D} < 0 \qquad \text{for } 1 - 2Q < A < 1 + 2Q \,,$$
$$\Delta \mathcal{D} = 0 \qquad \text{for } A = 1 \mp 2Q \,,$$
$$\Delta \mathcal{D} > 0 \qquad \text{for } A < 1 - 2Q \text{ or } A > 1 + 2Q \,,$$

which constitute property 1 of the list preceding (4.47).
When $\Delta \mathcal{D} < 0$, the solutions of $\mathcal{D} = 0$ are complex conjugates:

$$s_1,\ s_2 = -\sigma_n(A - 1) \pm j\sqrt{\omega_n^2 - \sigma_p^2(A - 1)^2} \,. \tag{4.52}$$

This is an alternative form of (4.48), obtained after making the square root real by changing the signs inside. The square distance R^2 of the poles from the origin is

$$R^2 = [\Re(s)]^2 + [\Im(s)]^2$$
$$= \sigma_n^2(A - 1)^2 + \omega_n^2 - \sigma_p^2(A - 1)^2 \,,$$

which simplifies to

$$R^2 = \omega_n^2 . \tag{4.53}$$

This is property 2 of the list preceding (4.47).

Finally, the poles are imaginary conjugates for $A = 1$. This is obtained by setting $A = 1$ in (4.52). The discriminant

$$\Delta \mathcal{D} = \sigma_p^2 (A - 1)^2 - \omega_n^2 \tag{4.54}$$

reduces to $-\omega_n^2$, and the real part $-\sigma_n(A - 1)$ of the solutions s_1, s_2 vanishes. Thus $s_1, s_2 = \pm j\omega_n$, which is property 3 of the above-mentioned list.

4.3.2　Further remarks

About the amplifier gain

Only the case $A = 1$ is relevant for oscillator design and operation. Nonetheless, the analysis of other cases provides insight.

> $A < 0$ corresponds to negative feedback. The resonator poles are pushed in the left-hand direction by the feedback, or made real for strong negative feedback $(A \leq 1 - 2Q)$.
>
> $A = 0$ corresponds to open-loop operation. The zeros and poles of $H(s)$ cancel one another, and the transfer function degenerates to $H(s) = 1$. This is consistent with the choice of the output point (Fig. 4.6(a)).
>
> $0 < A < 1$ corresponds to weak positive feedback. The resonator poles are pulled towards the imaginary axis without reaching it. The effect of the feedback is to sharpen the resonator's frequency response, yet without stationary oscillation.
>
> $A = 1$ corresponds to part of the Barkhausen condition for stationary oscillation. The poles are on the imaginary axis.
>
> $A > 1$ corresponds to strong positive feedback, which makes the oscillation amplitude diverge exponentially. The poles are on the right-hand half-plane, $\sigma > 0$. For $A > 1 + 2Q$ the positive feedback is so strong that the poles become real. In this case, the output rises exponentially with no oscillation.

Amplitude noise and frequency stability

That the poles of $H(s)$ are on a circle centered at the origin has an important consequence in metrology. In the real world *the gain fluctuates* around the value $A = 1$. For small fluctuations in A, the poles fluctuate perpendicularly to the imaginary axis. The effect on the oscillation frequency is second order only.

Gain control

The exact condition $A = 1$ cannot be ensured without a gain control mechanism. The latter can be interpreted as a control that stabilizes the oscillator poles on the imaginary axis.

The gain control can be a separate feedback system that sets A for the oscillator output voltage to be constant. This approach was implemented in early days of electronics by the Wien bridge oscillator [48]. Amplifier saturation, however, proved to be an effective amplitude control even in ultra-stable oscillators. When the amplifier saturates, the output signal is compressed more or less smoothly. The resultant power leakage from the fundamental to the harmonics reduces the gain and in turn stabilizes the output amplitude. The resonator prevents the harmonics from being fed back to the amplifier input.

Enhanced-quality-factor resonator

Positive feedback can be regarded as a trick to increase the quality factor Q of a resonator. This can be seen by equating the denominator

$$D(s) = s^2 + 2\sigma_p(A - 1)s + \omega_n^2$$

of the oscillator transfer function $H(s)$, (4.43), to the denominator

$$D_{eq} = s^2 + \frac{\omega_n}{Q_{eq}} s + \omega_n^2$$

of the transfer function $\beta_{eq}(s)$ of an equivalent resonator. The comparison gives

$$\frac{\omega_n}{Q_{eq}} = 2\sigma_p(A - 1) ; \tag{4.55}$$

thus, using $\sigma_p = -\omega_n/(2Q)$,

$$Q_{eq} = \frac{Q}{1 - A} . \tag{4.56}$$

The same conclusion can be drawn qualitatively by comparing Fig. 4.4 with Fig. 4.7, remembering that the complex conjugate poles are on a circle of radius ω_n.

Unfortunately, positive feedback cannot be taken as a way of escaping from the Leeson effect by artificially increasing the quality factor. In fact the noise reduction achieved in this way vanishes because of the additional noise introduced by the amplifier used to increase the quality factor, unless a superior technology is available.

4.4   Resonator in phase space

We will analyze the resonator phase response $b(t)$ and its Laplace transform $B(s)$ in quasi-stationary conditions. The resonator is driven by a sinusoidal signal at a frequency ω_0, which can be the resonator's natural frequency ω_n or any other frequency in a reasonable interval around ω_n. In the time domain the phase transfer function $b(t)$ is the phase of the resonator's response to a Dirac $\delta(t)$ function in the phase of the input. More precisely, $b(t)$ is defined as follows (Fig. 4.8(a)). Let

$$v_i(t) = \frac{1}{\beta_0} \cos(\omega_0 t - \theta) \quad \text{(stationary input)} \tag{4.57}$$

(a) Impulse response

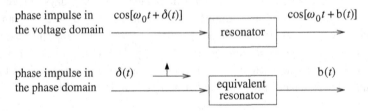

phase impulse in
the voltage domain

phase impulse in
the phase domain

(b) Step response

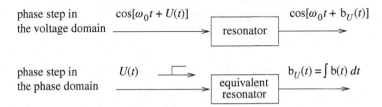

phase step in
the voltage domain

phase step in
the phase domain

Figure 4.8 Resonator phase response to a phase impulse $\delta(t)$.

be a stationary dimensionless[3] signal. The constants

$$\beta_0 = |\beta(j\omega_0)| \,, \tag{4.58}$$

$$\theta = \arg \beta(j\omega_0) \,, \tag{4.59}$$

are chosen so that the resonator's stationary output has unit amplitude and zero phase,

$$v_0(t) = \cos \omega_0 t \quad \text{(stationary output)}, \tag{4.60}$$

when the stationary signal (4.57) is fed into the input. Then, introducing an impulse $\delta(t)$ in the argument of the input,

$$v_i(t) = \frac{1}{\beta_0} \cos[\omega_0 t - \theta + \delta(t)] \quad \text{(input impulse)}, \tag{4.61}$$

we get the output transient

$$v_0(t) = \cos[\omega_0 t + b(t)] \,, \tag{4.62}$$

which defines $b(t)$. Of course, the impulse response $b(t)$ is obtained after linearizing the system. The resonator is described by a linear differential equation since the output is, necessarily, a linear function of the input. Yet understanding the process, and the meaning of this linearization, requires some simple mathematics, which we now introduce.

---

[3] The system function does not depend on the physical dimension of the input. Thus, we use dimensionless signals because this makes the formulae a little more concise.

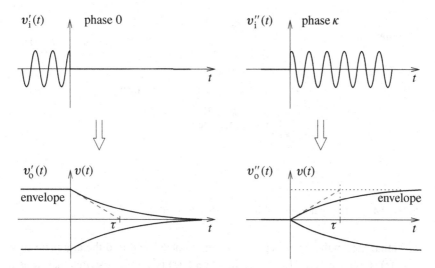

Figure 4.9 A sinusoid with a phase step can be decomposed into a sinusoid that is switched off at $t = 0$ plus a shifted sinusoid that is switched on at $t = 0$. The step response is the linear superposition of the two responses.

4.4.1   Phase-step method

The simplest way to find $b(t)$ is to feed a small phase step $\kappa U(t)$ into the argument of the input signal instead of the impulse $\delta(t)$. This is shown in Fig. 4.8(b) for $\kappa = 1$. The function

$$U(t) = \int_{-\infty}^{\infty} \delta(t)\,dt = \begin{cases} 0, & t < 0 \\ 1, & t > 0 \end{cases} \qquad \text{(Heaviside function)} \qquad (4.63)$$

is the well-known Heaviside function, also called the unit-step function. The impulse response $b(t)$ is obtained from the step response $b_U(t)$ using the property of linear systems that the impulse response is the derivative of the step response

$$b(t) = \frac{d}{dt}\, b_U(t)\,. \qquad (4.64)$$

That linearization is obtained for $\kappa \to 0$ is physically correct because in actual oscillators the phase noise is a small signal.

The phase-step method, shown in Fig. 4.9, consists of decomposing the input sinusoid

$$v_i(t) = \frac{1}{\beta_0} \cos\left[\omega_0 t - \theta + \kappa U(t)\right] \qquad (4.65)$$

into two truncated waveforms,

$$v_i(t) = v_i'(t) + v_i''(t)$$

$$= \frac{1}{\beta_0} \cos(\omega_0 t - \theta)\, U(-t) + \frac{1}{\beta_0} \cos(\omega_0 t - \theta + \kappa)\, U(t), \qquad (4.66)$$

$\underbrace{\phantom{\frac{1}{\beta_0} \cos(\omega_0 t - \theta)\, U(-t)}}_{v_i'(t),\ \text{switched off at } t = 0} \qquad \underbrace{\phantom{\frac{1}{\beta_0} \cos(\omega_0 t - \theta + \kappa)\, U(t)}}_{v_i''(t),\ \text{switched on at } t = 0}$

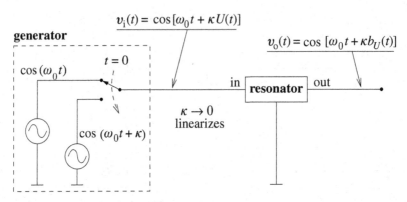

Figure 4.10 Electrical model of the phase-step method.

so that the small phase step $\kappa U(t)$ can be introduced into the argument of $v_i''(t)$ when $v_i''(t)$ starts. Here, the Heaviside function $U(t)$ is used as a switch that turns from off to on at $t = 0$, as shown in Fig. 4.10. Similarly, $U(-t)$ switches from on to off at $t = 0$. The output is

$$v_o(t) = v_o'(t) + v_o''(t) \qquad (\text{starting at } t = 0) , \tag{4.67}$$

where the terms $v_o'(t)$ and $v_o''(t)$ are, respectively, the switch-on and switch-off transient responses for $t \geq 0$. The output signal for $t < 0$ has no meaning in impulse-response analysis.

Before going through the analysis of the resonator response, it should be pointed out that the phase-step method finds application in the following investigation approaches.

Analytic method. This enables the calculation of the phase impulse response, as detailed in the following subsections.

Simulation tool. A driving sinusoid with a phase step can be regarded as the circuit shown in Fig. 4.10. This circuit is easy to simulate with any circuit-oriented simulation program, the most popular of which is Spice.

Experimental technique. A phase step can be implemented with a phase modulator driven by a low-frequency square wave. This can be a viable method for testing the resonator inside an oscillator. For example, in a coupled optoelectronic oscillator [109], where the microwave resonance results from the interaction between a microwave loop and an optical loop, the resonator cannot be separated from the oscillator. Nonetheless, the phase-step method has proved to be useful for measuring the closed-loop quality factor.

4.4.2   Input signal tuned exactly to the resonator's natural frequency

In this section we assume that the input frequency ω_0 is tuned to the exact natural frequency ω_n of the resonator:

$$\omega_0 = \omega_n . \tag{4.68}$$

We want to prove that the resonator impulse response in phase space is

$$b(t) = \frac{1}{\tau} e^{-t/\tau} \qquad \text{(impulse response)} \qquad (4.69)$$

$$= \omega_L e^{-\omega_L t}. \qquad (4.70)$$

That accomplished, the transfer function $B(s) = \mathcal{L}\{b(t)\}$ can be found in Laplace transform tables to be

$$B(s) = \frac{\omega_L}{s + \omega_L} \qquad \text{(transfer function)} \qquad (4.71)$$

$$= \frac{1/\tau}{s + 1/\tau}, \qquad (4.72)$$

which is a low-pass function. Additionally, it holds that

$$|B(j\omega)|^2 = \frac{1}{1 + \omega^2/\omega_L^2} = \frac{1/\tau^2}{\omega^2 + 1/\tau^2}. \qquad (4.73)$$

In order to prove (4.69), we first observe that $\beta(j\omega_0) = 1$; thus $\beta_0 = 1$ and $\theta = 0$. The input signal is

$$v_i(t) = \cos\left[\omega_0 t + \kappa U(t)\right]$$

$$= \cos \omega_0 t \, U(-t) + \cos(\omega_0 t + \kappa) \, U(t).$$

As in this section ω_0 and ω_n are taken to be equal, they may be used interchangeably. The resonator response $v_o'(t)$ to the sinusoid switched off at $t = 0$ is

$$v_o(t) = \begin{cases} \cos \omega_n t, & t \le 0, \\ \cos \omega_p t \, e^{-t/\tau}, & t > 0, \end{cases}$$

where $\tau = 2Q/\omega_n = 1/\omega_L$ is the resonator's relaxation time and $\omega_p = \omega_n\sqrt{1 - 1/(4Q^2)}$ is the free-decay pseudofrequency. Similarly, the response $v_o''(t)$ to the switched-on sinusoid is the exponentially growing sinusoid

$$v_o''(t) = \cos(\omega_p t + \kappa)\left(1 - e^{-t/\tau}\right), \qquad t > 0.$$

For $Q \gg 1$, we can make the approximation $\omega_p \simeq \omega_n = \omega_0$. This is justified by the fact that the phase error ζ accumulated during the relaxation time τ is

$$\zeta = (\omega_n - \omega_p)\tau = \frac{1}{4Q}.$$

This is seen by substituting $\tau = 2Q/\omega_n$ and $\omega_p = \omega_n\sqrt{1 - 1/(4Q^2)}$ into ζ, and by expanding in a series truncated at the first order for $Q \gg \frac{1}{2}$.

By virtue of linearity, the total output signal is

$$v_o(t) = v_o'(t) + v_o''(t), \qquad t > 0,$$

$$= \cos \omega_n t \, e^{-t/\tau} + (\cos \omega_n t \, \cos \kappa - \sin \omega_n t \, \sin \kappa)\left(1 - e^{-t/\tau}\right)$$

$$= \cos \omega_n t \left(e^{-t/\tau} + \cos \kappa - \cos \kappa \, e^{-t/\tau}\right) - \sin \omega_n t \, \sin \kappa \left(1 - e^{-t/\tau}\right).$$

For $\kappa \to 0$ we use the approximations $\cos \kappa \simeq 1$ and $\sin \kappa \simeq \kappa$. Thus

$$v_0(t) = \cos \omega_n t - \kappa \sin \omega_n t \left(1 - e^{-t/\tau}\right).$$

After factorizing out the time dependence $\omega_n t$, the above can be seen as a slowly varying phasor,

$$V_0(t) = \frac{1}{\sqrt{2}} \left[1 + j\kappa \left(1 - e^{-t/\tau}\right)\right], \qquad \kappa \ll 1.$$

The angle $\arctan \left(\Im\{V_0(t)\}/\Re\{V_0(t)\}\right)$, normalized on κ, is the step response

$$b_U(t) = 1 - e^{-t/\tau} \qquad \text{(step response)}.$$

The derivative of $b_U(t)$ is the impulse response $b(t) = (1/\tau)e^{-t/\tau}$, which is (4.69).

4.4.3   ★ Detuned input signal

Our aim is now to extend the results of the previous section to the general case where the input frequency ω_0 is not equal to the resonator's natural frequency ω_n. We want to prove that the impulse response of the resonator is

$$b(t) = \left(\Omega \sin \Omega t + \frac{1}{\tau} \cos \Omega t\right) e^{-t/\tau} \qquad \text{(impulse response, Fig. 4.12)} \qquad (4.74)$$

$$= \left(\Omega \sin \Omega t + \omega_L \cos \Omega t\right) e^{-\omega_L t}, \qquad (4.75)$$

that the transfer function is

$$B(s) = \frac{1}{\tau} \frac{s + 1/\tau + \Omega^2 \tau}{(s + 1/\tau - j\Omega)(s + 1/\tau + j\Omega)} \qquad \text{(transfer function, Fig. 4.13)}, \qquad (4.76)$$

and that

$$|B(j\omega)|^2 = \frac{1}{\tau^2} \frac{\omega^2 + \tau^2 \left(\Omega^2 + 1/\tau^2\right)^2}{\omega^4 - 2\left(\Omega^2 - 1/\tau^2\right)\omega^2 + \left(\Omega^2 + 1/\tau^2\right)^2} \qquad \text{(Fig. 4.14)} \qquad (4.77)$$

$$= \frac{\omega^2/\omega_L^2 + \left(1 + \Omega^2/\omega_L^2\right)^2}{\omega^4/\omega_L^4 + 2\left(\omega^2/\omega_L^2\right)\left(1 - \Omega^2/\omega_L^2\right) + \left(1 + \Omega^2/\omega_L^2\right)^2}, \qquad (4.78)$$

where the frequency offset, or detuning, Ω is defined as

$$\Omega = \omega_0 - \omega_n \qquad \text{(definition of detuning, } \Omega\text{)}. \qquad (4.79)$$

For reference, in the resonator bandwidth the detuning Ω spans the interval $-\omega_L$ to $+\omega_L$.

Preliminaries
Before introducing the phase step, we will study the transient of a resonator driven at the frequency $\omega_0 \neq \omega_n$.

Using the test signal

$$v_i(t) = \frac{1}{\beta_0} \cos(\omega_0 t - \theta) U(-t) , \qquad (4.80)$$

switched off at time $t = 0$, the output is

$$v_0(t) = \begin{cases} \cos \omega_0 t , & t \le 0 \\ \cos \omega_p t \, e^{-t/\tau} & t > 0 . \end{cases} \qquad (4.81)$$

For $t \le 0$ the output is determined by the choice of β_0 and θ, while for $t > 0$ the output is free exponential decay that is independent of ω_0.

Using the test signal

$$v_i(t) = \frac{1}{\beta_0} \cos(\omega_0 t - \theta) U(t) , \qquad (4.82)$$

switched on at time $t = 0$, the output is (see (4.20)),

$$v_0(t) = \mathscr{A} \cos \omega_p t \, e^{-t/\tau} + \mathscr{B} \sin \omega_p t \, e^{-t/\tau} + \mathscr{C} \cos \omega_0 t + \mathscr{D} \sin \omega_0 t \qquad t > 0 ,$$

where $\mathscr{A}$, $\mathscr{B}$, $\mathscr{C}$, and $\mathscr{D}$ are constants determined as follows. Given ω_0, the system has four degrees of freedom: the amplitude and phase of $v_i(t)$, the resonant frequency ω_n, and the quality factor Q. Thus the four unknowns $\mathscr{A}$, $\mathscr{B}$, $\mathscr{C}$, and $\mathscr{D}$ are completely determined. After our choice of input amplitude and phase, the output for $t \to \infty$ is $v_0(t) = \cos \omega_0 t$. This yields $\mathscr{C} = 1$ and $\mathscr{D} = 0$. Then $\mathscr{A}$ and $\mathscr{B}$ are found by using the continuity of the output signal at $t = 0$. This continuity condition gives $\mathscr{A} = -1$ and $\mathscr{B} = 0$. In summary,

$$v_0(t) = - \cos \omega_p t \, e^{-t/\tau} + \cos \omega_0 t , \qquad t > 0 . \qquad (4.83)$$

For $Q \gg 1$, we make the approximation $\omega_p \simeq \omega_n$. This is justified by the fact that the phase error accumulated during the relaxation time, $\zeta = (\omega_n - \omega_p)\tau = 1/(4Q)$, is small for large Q. Consequently, from (4.81) and (4.83) the output transients are

$$v_0(t) = \cos \omega_n t \, e^{-t/\tau} , \qquad\qquad t \le 0 \qquad \text{(switch-off)} , \qquad (4.84)$$
$$v_0(t) = - \cos \omega_n t \, e^{-t\tau} + \cos \omega_0 t \qquad t > 0 \qquad \text{(switch-on)} . \qquad (4.85)$$

Focusing on the switch-on transient, we have the following input–output relationship:

$$\begin{cases} v_i(t) = \dfrac{1}{\beta_0} \cos(\omega_0 t - \theta) U(t) , \\ v_0(t) = (- \cos \omega_n t \, e^{-t/\tau} + \cos \omega_0 t) U(t) \end{cases} \qquad (4.86)$$

and, as an obvious extension,

$$\begin{cases} v_i(t) = \dfrac{1}{\beta_0} \sin(\omega_0 t - \theta) U(t) , \\ v_0(t) = (- \sin \omega_n t \, e^{-t/\tau} + \sin \omega_0 t) U(t) . \end{cases} \qquad (4.87)$$

Introducing the phase step κ at t = 0

We use the test signal (4.66), here repeated:

$$v_i(t) = \underbrace{\frac{1}{\beta_0} \cos(\omega_0 t - \theta)\, U(-t)}_{v_i'(t),\ \text{switched off at } t = 0} + \underbrace{\frac{1}{\beta_0} \cos(\omega_0 t - \theta + \kappa)\, U(t)}_{v_i''(t),\ \text{switched on at } t = 0} \qquad (4.66)\,.$$

The input $v_i''(t)$ can be rewritten as

$$v_i''(t) = \frac{1}{\beta_0}\left[\cos(\omega_0 t - \varphi)\cos\kappa - \sin(\omega_0 t - \varphi)\sin\kappa\right] U(t)$$

$$= \frac{1}{\beta_0}\left[\cos(\omega_0 t - \varphi) - \kappa \sin(\omega_0 t - \varphi)\right] U(t) \qquad \text{for } \kappa \ll 1\,.$$

Using (4.84) and the pairs (4.86) and (4.87) we get for the output signal

$$v_o(t) = v_o'(t) + v_o''(t) \qquad \text{for } t > 0,$$

$$= \cos\omega_n t\, e^{-t/\tau} \qquad\qquad (\text{response to } v_i'(t))$$

$$+ (-\cos\omega_n t\, e^{-t/\tau} + \cos\omega_0 t) \qquad (\text{response to } v_i''(t), \text{ first part})$$

$$+ \kappa(\sin\omega_n t\, e^{-t/\tau} - \sin\omega_0 t) \qquad (\text{response to } v_i''(t), \text{ second part})\,.$$

Hence

$$v_o(t) = \cos\omega_0 t - \kappa \sin\omega_0 t + \kappa \sin\omega_n t\, e^{-t/\tau}, \qquad t > 0\,. \qquad (4.88)$$

Having defined the detuning frequency Ω as $\omega_0 - \omega_n$, (4.79), it holds that $\sin\omega_n t = \sin(\omega_0 t - \Omega t)$ and consequently that

$$\sin\omega_n t = \sin\omega_0 t \cos\Omega t - \cos\omega_0 t \sin\Omega t\,.$$

The output signal (4.88), rewritten in terms of ω_0 and Ω, is

$$v_o(t) = \cos\omega_0 t - \kappa \sin\omega_0 t + \kappa \sin\omega_0 t \cos\Omega t\, e^{-t/\tau} - \kappa \cos\omega_0 t \sin\Omega t\, e^{-t/\tau}\,,$$

which simplifies to

$$v_o(t) = \cos\omega_0 t(1 - \kappa \sin\Omega t\, e^{-t/\tau}) - \kappa \sin\omega_0 t(1 - \cos\Omega t\, e^{-t/\tau})\,. \qquad (4.89)$$

Phase impulse response b(t)

Freezing the oscillation $\omega_0 t$ (i.e. factoring it out), the output signal (4.89) turns into the slow-varying phasor

$$\mathbf{V}_o(t) = \frac{1}{\sqrt{2}}\left[1 + j\kappa(1 - \cos\Omega t\, e^{-t/\tau})\right], \qquad \kappa \ll 1\,.$$

The angle $\arctan(\Im\{V_o(t)\}/\Re\{V_o(t)\})$, normalized to κ, is the phase-step response

$$b_U(t) = 1 - \cos\Omega t\, e^{-t/\tau} \qquad (\text{step response, Fig. 4.11})\,.$$

Using the property $b(t) = (d/dt)\, b_U(t)$, (4.64), we find the impulse response

$$b(t) = \left(\Omega \sin\Omega t + \frac{1}{\tau}\cos\Omega t\right) e^{-t/\tau} \qquad (\text{impulse response, Fig. 4.12})\,.$$

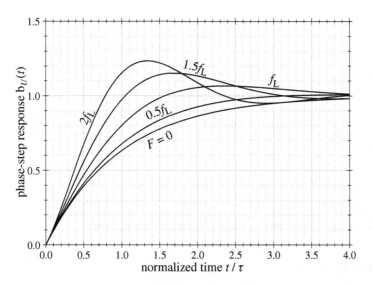

Figure 4.11 Resonator phase-step response $b_U(t)$; $F = \Omega/(2\pi)$, see (4.79).

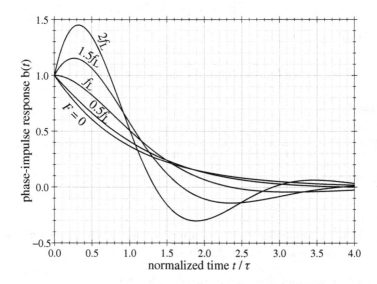

Figure 4.12 Resonator phase-impulse response $b(t)$; $F = \Omega/(2\pi)$, see (4.79).

Phase frequency response $B(s)$

The Laplace transform $B(s) = \mathcal{L}\{b(t)\}$ is found using the Euler formulae

$$\cos \Omega t = \frac{1}{2}\left(e^{j\Omega t} + e^{-j\Omega t}\right),$$

$$\sin \Omega t = \frac{1}{j2}\left(e^{j\Omega t} - e^{-j\Omega t}\right)$$

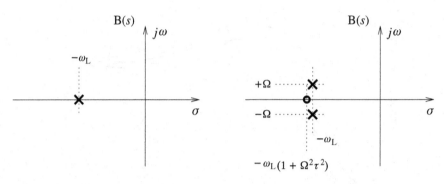

Figure 4.13 Phase-noise transfer function $B(s)$ on the complex plane.

and the properties

$$\mathcal{L}\{e^{-t/\tau}\} = \frac{1}{s + 1/\tau},$$

$$\mathcal{L}\{e^{at} f(t)\} = F(s - a).$$

Expanding $\mathcal{L}\{b(t)\}$, we find

$$B(s) = \mathcal{L}\left\{ \left[\Omega \frac{1}{j2} \left(e^{j\Omega t} - e^{-j\Omega t} \right) + \frac{1}{\tau}\frac{1}{2} \left(e^{j\Omega t} + e^{-j\Omega t} \right) \right] e^{-t/\tau} \right\},$$

and finally

$$B(s) = \frac{1}{\tau} \frac{s + 1/\tau + \Omega^2 \tau}{(s + 1/\tau - j\Omega)(s + 1/\tau + j\Omega)} \qquad \text{(transfer function, Fig. 4.13)},$$

which is (4.76).

Remark.
The phase-noise bandwidth of the resonator increases when the resonator is detuned (Fig. 4.14). This is related to the following facts.

1. When the resonator is detuned, it holds that (Fig. 4.4)

$$\left| \frac{d \arg \beta(j\omega)}{d\omega} \right|_{\omega_0} < \left| \frac{d \arg \beta(j\omega)}{d\omega} \right|_{\omega_n}. \qquad (4.90)$$

For a lower slope, the oscillator phase noise is higher.
2. Detuning the resonator, the symmetry of $\arg \beta(j\omega)$ around the oscillation frequency is lost. This explains the overshoot seen in Fig. 4.14 for $\Omega \neq 0$.
3. It may be seen in Fig. 4.11 that the step response is faster when the resonator is detuned.

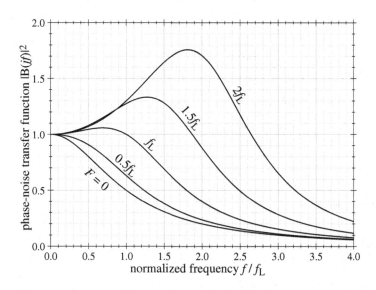

Figure 4.14 Modulus squared of the phase-noise transfer function, $|\mathrm{B}(j\omega)|^2$, (4.76); $F = \Omega/(2\pi)$.

4.5   Proof of the Leeson formula

Owing to the practical need for stabilizing the output amplitude, the oscillator is in-herently nonlinear. Yet, a linear model is correct for small phase perturbations to the stationary oscillation. This statement can be justified further if the noisy oscillator is described by a stochastic equation whose solution converges on the limit cycle. In the case of a sinusoidal oscillation, the limit cycle is a circle or an ellipse. Of course, a smoothly nonlinear system can be linearized for small perturbations around the limit cycle. In actual oscillators the phase noise is always a small perturbation.

That said, we will abandon the generic model of Fig. 4.15(a) in favor of the model of Fig. 4.15(b), which is specific to phase noise. This model describes phase perturbations around a stationary oscillation, for which $A\beta(j\omega) = 1$ at the oscillation frequency ω_0. The physical quantities of Fig. 4.15(b) are the Laplace transforms of the oscillator's phase fluctuations. Thus the input signal $\Psi(s)$ models the amplifier's phase noise, or other noise sources referred to the amplifier input. Additionally, $\Psi(s)$ can be used to introduce resonator fluctuations, or the phase noise of an external signal to which the oscillator is injection-locked. In phase space, the amplifier gain is exactly unity because the amplifier repeats the input phase to the output. The feedback function is B(s), discussed thoroughly in Section 4.4 above.

Describing phase noise, the oscillator transfer function is defined as

$$H(s) = \frac{\Phi(s)}{\Psi(s)} \qquad \text{(definition of H(s))}, \tag{4.91}$$

(a) Voltage space

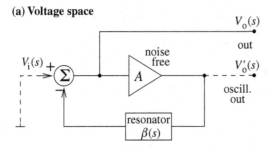

(b) Phase space

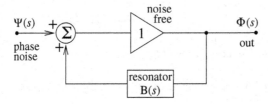

Figure 4.15 Derivation of the oscillator phase-noise model (b) from the voltage-noise oscillator model (a).

and thus

$$S_\varphi(\omega) = |H(j\omega)|^2 \, S_\psi(\omega). \tag{4.92}$$

From the general feedback theory, with the scheme of Fig. 4.15(b) the transfer function is given by

$$H(s) = \frac{1}{1 - B(s)}. \tag{4.93}$$

Hence the function $H(s)$ is the essence of the Leeson formula (3.19).

4.5.1   Oscillator tuned exactly to the resonator's natural frequency

At the exact natural frequency ω_n, the resonator's phase transfer function is $B(s) = 1/(s\tau + 1)$, (4.71). Substituting $B(s)$ into (4.93) we get

$$H(s) = \frac{1 + s\tau}{s\tau} \qquad \text{(Fig. 4.16(a))} ; \tag{4.94}$$

thus

$$|H(j\omega)|^2 = \frac{\tau^2\omega^2 + 1}{\tau^2\omega^2} \qquad \text{(Fig. 4.17)} \tag{4.95}$$

$$= \frac{1 + \omega^2/\omega_L^2}{\omega^2/\omega_L^2}. \tag{4.96}$$

Combining (4.92) and (4.95),

$$S_\varphi(\omega) = \left(1 + \frac{1}{\omega^2\tau^2}\right) S_\psi(\omega) \tag{4.97}$$

$$= \left(1 + \frac{\omega_L^2}{\omega^2}\right) S_\psi(\omega) , \tag{4.98}$$

recalling that $\tau = 2Q/\omega_0 = Q/(\pi v_0) = 1/\omega_L$, (4.18), and using $\omega = 2\pi f$ we get

$$S_\varphi(f) = \left(1 + \frac{1}{4\pi^2 f^2} \frac{\pi^2 v_0^2}{Q^2}\right) S_\psi(f) \tag{4.99}$$

and finally

$$S_\varphi(f) = \left[1 + \frac{1}{f^2} \left(\frac{v_0}{2Q}\right)^2\right] S_\psi(f) \qquad \text{(Leeson formula)} . \tag{4.100}$$

This confirms the Leeson formula (3.19), which was found in Chapter 3 using heuristic argumentation and physical insight.

4.5.2   ★ Detuned oscillator

Let us consider the more general case of an oscillator oscillating at the frequency $\omega_0 = \omega_n + \Omega$, with $\Omega \neq 0$. The phase-noise transfer function $H(s)$ is found by substituting $B(s)$, (4.76), into (4.93). Using the simple property that $1/(1 - N/D) = D/(D - N)$, we get

$$H(s) = \frac{\tau(s + 1/\tau - j\Omega)(s + 1/\tau + j\Omega)}{\tau(s + 1/\tau - j\Omega)(s + 1/\tau + j\Omega) - (s + 1/\tau + \Omega^2\tau)}$$

$$= \frac{(s\tau + 1 - j\Omega\tau)(s\tau + 1 + j\Omega\tau)}{(s\tau + 1 - j\Omega\tau)(s\tau + 1 + j\Omega\tau) - (s\tau + 1 + \Omega^2\tau^2)}$$

$$= \frac{(s\tau + 1)^2 + \Omega^2\tau^2}{(s\tau + 1)^2 + \Omega^2\tau^2 - (s\tau + 1) - \Omega^2\tau^2} ,$$

and hence

$$H(s) = \frac{(s\tau + 1)^2 + \Omega^2\tau^2}{s\tau(s\tau + 1)} \qquad \text{(detuned by Ω)} , \tag{4.101}$$

or

$$H(s) = \frac{(s + 1/\tau - j\Omega)(s + 1/\tau + j\Omega)}{s\tau(s + 1/\tau)} \qquad \text{(alternative form of (4.101))} , \tag{4.102}$$

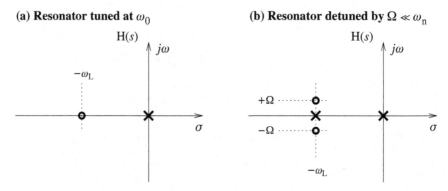

(a) Resonator tuned at ω_0   **(b) Resonator detuned by $\Omega \ll \omega_n$**

Figure 4.16 Oscillator phase-noise transfer function H(s) on the complex plane.

and so

$$|H(j\omega)|^2 = \frac{\tau^4\omega^4 + 2(\tau^2 - \Omega^2\tau^4)\omega^2 + (\Omega^2\tau^2 + 1)^2}{\tau^2\omega^2(\tau^2\omega^2 + 1)} \tag{4.103}$$

$$= \frac{\omega^4/\omega_L^4 + 2\left(1 + \Omega^2/\omega_L^2\right)\left(\omega^2/\omega_L^2\right) + \left(1 + \Omega^2/\omega_L^2\right)^2}{\left(\omega^2/\omega_L^2\right)\left(1 + \omega^2/\omega_L^2\right)}. \tag{4.104}$$

The derivation of $|H(j\omega)|^2$ is omitted.

Figure 4.16 shows H(s) in the complex plane. When the oscillator is pulled away from ω_n, the real zero at $s = -\omega_L$ splits into a pair of complex conjugate zeros at $s = -\omega_L \pm j\Omega$, leaving a real pole at $s = -\omega_L$ in between. The pole at $s = 0$ in Fig. 4.16 is an ideal integrator in the time domain. This leads to the Leeson effect, i.e. the divergence of the phase noise in the long run. Seen from the imaginary axis at $\omega \gg \omega_L$, H(s) appears as a small cluster of two poles and two zeros that null one another, for $H(j\omega)$ is constant. The resonator is a flywheel that blocks the phase fluctuations, since the oscillator is an open loop. Accordingly, the amplifier phase noise $\Psi(s)$ is repeated as the output. This is the term "1" in the Leeson formula (4.100).

The plot of $|H(j\omega)|^2$, shown in Fig. 4.17, reveals that the phase-noise response is a function of Ω that increases as Ω increases. This is best shown by the integral

$$\int_0^\infty \frac{\left|H(j\omega)\right|^2 - \left|H(j\omega)\right|^2_{\Omega=0}}{\left|H(j\omega)\right|^2_{\Omega=0}} \, d\omega = \frac{\pi}{4} \Omega^4\tau^3. \tag{4.105}$$

4.5.3   Pulling the oscillator frequency

The oscillator can be pulled to a desired frequency in an interval around ω_n, as shown in subsection 3.1.2.

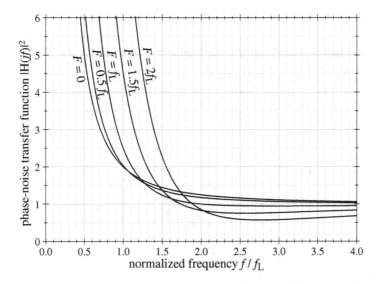

Figure 4.17 Modulus squared of the phase-noise transfer function $|H(j\omega)|^2$, (4.104); $F = \Omega/(2\pi)$.

If the oscillator is pulled by inserting a static phase into the loop, the oscillation frequency is not equal to the resonator's natural frequency. In this condition it holds that $\Omega \neq 0$. Equation (4.105) indicates that the oscillator phase noise then increases. However, if ω_n is changed by making the resonator interact with an external reactance then Ω is unaffected. Thus the phase noise does not increase. The following cases are of interest.

Quartz oscillator
The quartz oscillator is usually pulled by means of an external reactance, as in Fig. 3.7, for Ω is unchanged. Yet the additional loss introduced by the external reactance may lower the quality factor and in turn increase the resonator noise bandwidth.

Low-noise microwave oscillator
In low-noise schemes such as the Galani oscillator [34], it often happens that the resonator cannot be tuned, because a phase shifter is used to pull the oscillation frequency. It follows that $\Omega \neq 0$ and additional noise will be taken in.

Pound-stabilized oscillator
In the Pound scheme [77], the modulation and detection mechanisms ensure that the resonator is tuned at the exact natural frequency ω_n by equating the amplitudes of the reflected sidebands. Therefore in this oscillator it holds that $\Omega = 0$ unless a frequency offset is forced intentionally through the control loop.

4.6   Frequency-fluctuation spectrum and Allan variance

If the phase-noise spectrum inside the loop is $S_\psi(f) = b_0 + b_{-1}f^{-1} + \cdots$, the oscillator's fractional frequency spectrum is

$$S_y(f) = \frac{b_0}{v_0^2}f^2 + \frac{b_{-1}}{v_0^2}f + \frac{b_0}{4Q^2} + \frac{b_{-1}}{4Q^2}\frac{1}{f}. \tag{4.106}$$

This is easily seen as follows:

$$S_y(f) = \frac{f^2}{v_0^2}S_\varphi(f) \qquad \text{(derivative)}$$

$$= \frac{f^2}{v_0^2}\left(1 + \frac{1}{f^2}\frac{v_0^2}{4Q^2}\right)S_\psi(f) \qquad \text{(Leeson)}$$

$$= \left(\frac{f^2}{v_0^2} + \frac{1}{4Q^2}\right)S_\psi(f)$$

$$= \frac{b_0}{v_0^2}f^2 + \frac{b_{-1}}{v_0^2}f + \frac{b_0}{4Q^2} + \frac{b_{-1}}{4Q^2}\frac{1}{f}.$$

The proof extends this result to higher-order terms, in $b_{-2}f^{-2}$, etc.

The oscillator's Allan variance is

$$\sigma_y^2(\tau) = (1/\tau^2 \text{ terms}) + \frac{1}{2}\frac{1}{4Q^2}b_0\frac{1}{\tau} + 2\ln 2\frac{1}{4Q^2}b_{-1} + \cdots \tag{4.107}$$

This can be demonstrated by matching (4.106) to the power law $S_y(f) = \sum_i h_i f^i$, and by identifying the terms; thus

$$h_0 = \frac{1}{4Q^2}b_0 \qquad \text{and} \qquad h_{-1} = \frac{1}{4Q^2}b_{-1}.$$

The coefficients h_0 and h_{-1} are converted into the Allan variance using Table 1.4 in Section 1.8. It is worth mentioning that, for reasons detailed in Chapter 2, terms of higher order than $b_{-2}f^{-2}$ cannot be present in the amplifier noise. They can be included in the formula for the sake of completeness, however, because $\psi(t)$ models all the phase fluctuations present in the loop.

Example 4.1. Calculate the Allan variance and deviation of a microwave dielectric resonator oscillator (DRO) in which the resonator quality factor is $Q = 2500$ and the amplifier noise is $S_\varphi(f) = 10^{-15} + 10^{-11}/f$ (white noise -150 dB rad^2/Hz), which results from $F = 4$ dB, $P_0 = -20$ dB m, and flicker noise -110 dB rad^2/Hz at 1 Hz). Using (4.107),

$$\sigma_y^2(\tau) = \frac{2 \times 10^{-23}}{\tau} + 5.55 \times 10^{-19}, \qquad \sigma_y(\tau) \approx \frac{4.47 \times 10^{-12}}{\sqrt{\tau}} + 7.45 \times 10^{-10}.$$

4.7   ★★ A different, more general, derivation of the resonator phase response

The derivation of the resonator phase response[4] is approached here in a completely different way, which generalizes the results of Section 4.4 in that:

- the resonator equation is left arbitrary;
- it predicts the possibility of a phase–amplitude interaction.

Let us denote by $b(t)$ the resonator's (ordinary) impulse response and by $\beta(j\omega)$ its Fourier transform. Further, let us denote by $b(t)$ the resonator's phase-impulse response and by $B(j\omega)$ its Fourier transform. We are familiar with the functions $\beta(j\omega)$, $b(t)$, and $B(j\omega)$; $b(t)$ is introduced here for the first time. We will assume that the resonator is replaced by a band-pass filter whose frequency response is a sharp peak at $\omega \approx \omega_0$. In a band-pass filter, the system roots that produce a frequency-selective behavior are clustered in two symmetrical regions around $\pm j\omega_0$. Additionally, zeros are present at the origin and at infinity. Such a filter is a generalization of a $Q \gg 1$ resonator that can be used to model a variety of physical systems, the most interesting of which are a resonator having other resonances in the vicinity of the oscillation frequency and a resonator with quasi-degenerate or degenerate resonances at the oscillation frequency.

Letting

$$v_i(t) = \cos[\omega_0 t + \zeta + \varphi_i(t)] = \Re\left\{e^{j\omega_0 t}\, e^{j\zeta}\, e^{j\varphi_i(t)}\right\} \qquad \text{(input signal)} \qquad (4.108)$$

be the input signal, the output signal is then

$$v_o(t) = (b * v_i)(t) \qquad \text{(convolution)}$$

$$= \int_{-\infty}^{\infty} b(u)v_i(t - u)\, du$$

$$= \Re\left\{\int_{-\infty}^{\infty} b(u)\, e^{j\omega_0(t-u)} e^{j\zeta} e^{j\varphi_i(t-u)}\, du\right\}$$

$$= \Re\left\{e^{j\omega_0 t} e^{j\zeta} \int_{-\infty}^{\infty} b(u)e^{-j\omega_0 u} e^{j\varphi_i(t-u)}\, du\right\}.$$

Owing to the high signal-to-noise ratio of real oscillators, we can linearize the expression of $v_o(t)$ for $|\varphi_i(t)| \ll 1$:

$$v_o(t) = \Re\left\{e^{j\omega_0 t} e^{j\zeta} \int_{-\infty}^{\infty} b(u)e^{-j\omega_0 u}\left[1 + j\varphi_i(t - u)\right] du\right\}$$

$$= \Re\left\{e^{j\omega_0 t} e^{j\zeta}\left[\beta(j\omega_0) + j\int_{-\infty}^{\infty} b(u)e^{-j\omega_0 u}\varphi_i(t - u)\, du\right]\right\}.$$

---

[4] The analytical method used here was suggested by Charles Greenhall (private communication).

Making the replacement $\beta(j\omega_0) = \beta_0 e^{j\theta}$, we have

$$v_0(t) = \Re\left\{ e^{j\omega_0 t} e^{j\zeta} \left[\beta_0 e^{j\theta} + j \int_{-\infty}^{\infty} b(u) e^{-j\omega_0 u} \varphi_i(t-u)\, du \right] \right\}$$

$$= \Re\left\{ e^{j\omega_0 t} e^{j\zeta} \beta_0 e^{j\theta} \left[1 + \frac{j}{\beta_0} \int_{-\infty}^{\infty} b(u) e^{-j(\omega_0 u + \theta)} \varphi_i(t-u)\, du \right] \right\}.$$

Using $e^{jx} = \cos x + j \sin x$, we observe that

$$je^{-j(\omega_0 u + \theta)} = \sin(\omega_0 u + \theta) + j \cos(\omega_0 u + \theta).$$

Defining

$$b_c(t) = b(t) \cos(\omega_0 t + \theta),\tag{4.109}$$

$$b_s(t) = b(t) \sin(\omega_0 t + \theta),\tag{4.110}$$

we get

$$v_0(t) = \Re\left\{ e^{j\omega_0 t} e^{j\zeta} \beta_0 e^{j\theta} \left[1 + \frac{1}{\beta_0} \int_{-\infty}^{\infty} b_s(u)\, \varphi_i(t-u)\, du \right.\right.$$

$$\left.\left. + \frac{j}{\beta_0} \int_{-\infty}^{\infty} b_c(u)\, \varphi_i(t-u)\, du \right] \right\}$$

$$= \Re\left\{ e^{j\omega_0 t} e^{j\zeta} \beta_0 e^{j\theta} \left[1 + \frac{(b_s * \varphi_i)(t)}{\beta_0} + j\frac{(b_c * \varphi_i)(t)}{\beta_0} \right] \right\}.$$

Using $1 + x + jy \approx (1+x)e^{jy}$ for small x and y, we get

$$v_0(t) = \Re\left\{ e^{j\omega_0 t} e^{j\zeta} \beta_0 e^{j\theta} \left[1 + \frac{(b_s * \varphi_i)(t)}{\beta_0} \right] \exp\left[j\frac{(b_c * \varphi_i)(t)}{\beta_0} \right] \right\}.$$

We aim to identify the fractional amplitude $\alpha_0(t)$ and the phase $\varphi_0(t)$ at the resonator output, making an implicit reference to the output signal written as $v_0(t) = [1 + \alpha_0(t)] \cos[\omega_0 t + \varphi_0(t)]$, as in (1.6). For this purpose, it is useful to *normalize the input signal* so that $e^{j\zeta} \beta_0 e^{j\theta} = 1$. This means that we set $\zeta = -\theta$ and scale up the amplitude by a factor $1/\beta_0$. The output signal thereupon becomes

$$v_0(t) = \Re\left\{ e^{j\omega_0 t} \left[1 + \frac{(b_s * \varphi_i)(t)}{\beta_0} \right] \exp\left[j\frac{(b_c * \varphi_i)(t)}{\beta_0} \right] \right\}.$$

The output phase and amplitude are

$$\varphi_0(t) = \frac{1}{\beta_0} (b_c * \varphi_i)(t) \qquad \text{(PM} \rightarrow \text{PM conversion)},\tag{4.111}$$

$$\alpha_0(t) = \frac{1}{\beta_0} (b_s * \varphi_i)(t) \qquad \text{(PM} \rightarrow \text{AM conversion)}.\tag{4.112}$$

The filter functions $b_c(t)$ and $b_s(t)$ can be rewritten using $\cos x = \frac{1}{2}\left(e^{jx} + e^{-jx}\right)$ and $\sin x = \frac{1}{j2}\left(e^{jx} - e^{-jx}\right)$:

$$b_c(t) = \frac{1}{2}\left[e^{j(\omega_0 t + \theta)} + e^{-j(\omega_0 t + \theta)}\right]b(t),\qquad(4.113)$$

$$b_s(t) = \frac{1}{j2}\left[e^{j(\omega_0 t + \theta)} - e^{-j(\omega_0 t + \theta)}\right]b(t).\qquad(4.114)$$

Interestingly, (4.112) shows the existence of a phase-to-amplitude noise conversion, which may null in some specific conditions.

The expressions (4.111) and (4.112) yield naturally the transfer functions

$$B(j\omega) = \frac{\Phi_o(j\omega)}{\Phi_i(j\omega)}\qquad (\text{PM} \to \text{PM conversion}),\qquad(4.115)$$

$$A(j\omega) = \frac{A_o(j\omega)}{\Phi_i(j\omega)}\qquad (\text{PM} \to \text{AM conversion}),\qquad(4.116)$$

where upper case denotes a Fourier transform; $B(j\omega)$ is the phase transfer function with which we are familiar, $A_o(j\omega) = \mathcal{F}\{\alpha_o(t)\}$ is the normalized output amplitude, and $A(j\omega)$ is the phase-to-amplitude conversion function. Combining (4.111) and (4.113), we get

$$B(j\omega) = \mathcal{F}\left\{\frac{1}{\beta_0}\frac{1}{2}\left(e^{j\omega_0 t}e^{j\theta} + e^{-j\omega_0 t}e^{-j\theta}\right)b(t)\right\}$$

$$= \frac{1}{2\beta_0}\left\{\delta(j(\omega - \omega_0))e^{j\theta} + \delta(j(\omega + \omega_0))e^{-j\theta}\right\} * B(j\omega).$$

Similarly, combining (4.112) and (4.114), we get

$$A(j\omega) = \mathcal{F}\left\{\frac{1}{\beta_0}\frac{1}{j2}\left(e^{j\omega_0 t}e^{j\theta} - e^{-j\omega_0 t}e^{-j\theta}\right)b(t)\right\}$$

$$= \frac{1}{j2\beta_0}\left\{\delta(j(\omega - \omega_0))e^{j\theta} - \delta(j(\omega + \omega_0))e^{-j\theta}\right\} * B(j\omega).$$

Hence

$$B(j\omega) = \frac{1}{2\beta_0}\left\{B(j(\omega - \omega_0))e^{j\theta} + B(j(\omega + \omega_0))e^{-j\theta}\right\},\qquad(4.117)$$

$$A(j\omega) = \frac{1}{j2\beta_0}\left\{B(j(\omega - \omega_0))e^{j\theta} - B(j(\omega + \omega_0))e^{-j\theta}\right\}.\qquad(4.118)$$

The above equations (4.117) and (4.118) apply to the region around $\omega = 0$ where phase-noise mechanisms take place. The reason comes from the definitions of $b_c(t)$ and $b_s(t)$ (see (4.109) and (4.110)), which produce up-conversion and down-conversion of $b(t)$, respectively, because of the multiplication by a sinusoid. This is illustrated in Fig. 4.18, where the Fourier transform is replaced by the Laplace transform and shown in the complex plane. The region of the system function around $j\omega_0$ that contains the roots generating the band-pass behavior is up-converted to $2j\omega_0$ and down-converted to dc.

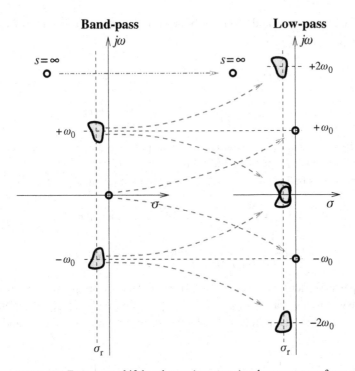

Figure 4.18 Frequency-shift band-pass (resonator) to low-pass transformation.

The same thing happens to the symmetrical region, around $-j\omega_0$. The roots of $A(s)$ and $B(s)$ in the regions around $\pm j\omega_0$ and $\pm 2j\omega_0$ correspond theoretically to phase noise at Fourier frequencies equal to the carrier frequency or twice the carrier frequency, but this has no practical relevance. Before getting into confusion about this point, one should recall that $A(s)$ and $B(s)$ refer to the phase behavior of the filter. The phase, though mathematically defined at any frequency, makes no sense far from the band-pass region because there is virtually no signal.

4.7.1 Input signal tuned exactly to the resonator's natural frequency

When the input signal is tuned exactly to the resonator's natural frequency, $\omega_0 = \omega_n$, it holds that $\theta = 0$ and hence $e^{\pm j\theta} = 1$. Owing to the symmetry of $\beta(j\omega)$ (subsection 4.2.1), (4.117) and (4.118) reduce to

$$B(j\omega) = \frac{1}{\beta_0}\beta(j(\omega - \omega_n)),$$
(4.119)

$$A(j\omega) = 0.$$
(4.120)

This is the result already found in subsection 4.4.2; the pole pair of $\beta(s)$ at $s = \sigma_p \pm j\omega_p$ turns into the pole of $B(s)$ at $s = \sigma_p$.

4.8   ★★ Frequency transformations

The classical theory of filter networks suggests that an arbitrary filter can be obtained from a low-pass prototype by conformal transformation of the complex variable [43, Section 14.4, 101, Section 13.4, 106, Section 11.9]. The transformation that is most interesting for us is $s \to s + 1/s$, because it turns a low-pass filter into a band-pass by adding a high-pass function obtained from the low-pass function using the transformation $s \to 1/s$. Here we prefer to show the analogy between band-pass (BP) and a "true" low-pass (LP), instead of using the dimensionless frequency of the LP prototype.

We start from the single-pole LP function

$$H_{LP}(p) = \frac{1/\tau}{p + 1/\tau} , \tag{4.121}$$

where the complex variable p is used instead of s to avoid confusion. By making the replacement

$$p = \frac{1}{2} \frac{s^2 + \omega_n^2}{s} \qquad \text{(LP} \to \text{BP transformation)} \tag{4.122}$$

in (4.121), we find that

$$\beta(s) = \frac{2}{\tau} \frac{s}{s^2 + 2s/\tau + \omega_n^2} . \tag{4.123}$$

Finally, using $\tau = 2Q/\omega_n$, thus $2/\tau = \omega_n/Q$, we see that (4.123) is equivalent to the canonical form (4.25) of the resonator, here repeated:

$$\beta(s) = \frac{\omega_n}{Q} \frac{s}{s^2 + \omega_n s/Q + \omega_n^2} \tag{4.25} .$$

It is instructive to learn more about how a frequency transformation maps an LP function into a BP function (Fig. 4.19). From (4.122) we get

$$s = p \pm \sqrt{p^2 - \omega_n^2} \tag{4.124}$$

$$= p \pm j\omega_n \sqrt{1 - p^2/\omega_n^2} . \tag{4.125}$$

Thus, we can deduce the following.

- The point $p = \infty$ maps into $s = 0$. This is seen by evaluating the limit of (4.124) for $p \to \infty$.
- The point $p = 0$ maps into $s = j\omega_n$. This is seen in (4.125).
- After expanding p as $\Sigma + j\Omega$, a point $p = \Sigma$ on the real axis maps into

$$s = \Sigma \pm \sqrt{\Sigma^2 - \omega_n^2} , \tag{4.126}$$

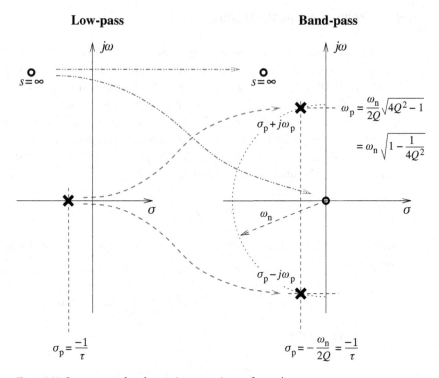

Figure 4.19 Low-pass to band-pass (resonator) transformation.

that is,

$$s = \Sigma \pm j\omega_n \sqrt{1 - \Sigma^2/\omega_n^2} \qquad \text{for } \Sigma < \omega_n , \tag{4.127}$$

$$s = \Sigma \pm \Sigma \sqrt{1 - \omega_n^2/\Sigma^2} \qquad \text{for } \Sigma > \omega_n . \tag{4.128}$$

- Similarly, a point $p = j\Omega$ on the imaginary axis maps into

$$s = -j\Omega \pm j\omega_n \sqrt{\Omega^2 + \omega_n^2} , \tag{4.129}$$

that is ,

$$s = j\Omega \pm j\omega_n \sqrt{1 + \Omega^2/\omega_n^2} \tag{4.130}$$

or

$$s = j\Omega \pm j\Omega \sqrt{1 + \omega_n^2/\Omega^2} . \tag{4.131}$$

Replacing s by $j\omega$ in the transformation (4.122), we get

$$p = \frac{1}{2} j \omega_n \left(\frac{\omega}{\omega_n} - \frac{\omega_n}{\omega} \right) \qquad ((4.122) \text{ with } s = j\omega) . \tag{4.132}$$

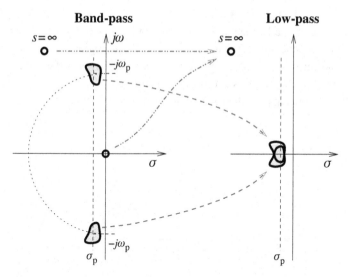

Figure 4.20 General band-pass (resonator) to low-pass transformation.

thus, using the definition (4.31) of the dissonance χ,

$$p = \frac{1}{2} j\omega_n \chi \ . \tag{4.133}$$

Dividing both sides of the above by $\omega_L = \omega_n/(2Q)$, and recalling that $\omega_L = 1/\tau$, we get

$$p\tau = jQ\chi \ , \tag{4.134}$$

which allows us to rewrite the expression (4.121) for the LP equivalent function:

$$H_{LP}(\chi) = \frac{1}{1 + jQ\chi} \ . \tag{4.135}$$

The natural extension of this is to replace the imaginary variable $j\chi$ by the *complex dissonance*

$$\mathfrak{X} = \xi + j\chi \ . \tag{4.136}$$

It is to be remarked that the dissonance χ, and therefore also the complex dissonance $\mathfrak{X}$, is a warped and normalized frequency and that it is dimensionless. Using $\mathfrak{X}$, the equivalent LP filter is

$$H_{LP}(\mathfrak{X}) = \frac{1}{1 + Q\mathfrak{X}} \ . \tag{4.137}$$

Finally, for the purposes of understanding the resonator we prefer the inverse transformation, that is, the transformation from band-pass to low-pass. A generalized form of this transformation is shown in Fig. 4.20. The system function has two symmetrical sets of roots generating the band-pass behavior, in the region around $j\omega_p$ and in the region around $-j\omega_p$. Additionally, the band-pass function has one or more zeros at $s = 0$ and at $s = \infty$. The frequency transformation moves the $\pm j\omega_p$ regions to the origin, overlapping them. The zeros at $s = \infty$ are preserved, while the zeros at the origin are moved to $s = \infty$. The problem with this conformal transformation is that it warps the complex plane, splitting or overlapping the roots; thus one loses insight.

5 Noise in delay-line oscillators and lasers

The basic delay-line oscillator, shown in Fig. 5.1, is an oscillator in which the feedback path has a delay τ_d that is independent of frequency. Hence, when the amplifier gain $A = 1$, the Barkhausen condition is met at any frequency $\omega_l = (2\pi/\tau_d)l$ with l integer, that is, a frequency multiple of the free spectral range $2\pi/\tau_d$. With appropriate initial conditions, stationary oscillation takes place. If the exact condition $A = 1$ is met in a frequency range, several oscillation frequencies may coexist.

A real oscillator requires a small-signal gain $A > 1$ that reduces to unity in appropriate large-signal conditions. Of course, a selector filter is necessary in order to choose a mode m and thus a frequency $\omega_m = (2\pi/\tau_d)m$. The filter introduces attenuation at all frequencies ω_l with $l \neq m$. For laboratory demonstration purposes, the imperfect gain flatness as a function of frequency of real amplifiers is sufficient to select a mode, albeit an arbitrary one. A true band-pass filter is necessary for practical applications. Using a tunable filter the oscillation frequency can be switched between modes, as in a synthesizer, in steps $\Delta\omega = 2\pi/\tau_d$.

It will be shown that the laser is a special case of a delay-line oscillator. Additionally, the delay-line oscillator is of great interest for the generation of microwaves and THz waves from optics because a long delay can be implemented thanks to the high transparency of some materials, such as silica, CaF_2, and MgF_2, at 1.55 μm wavelength. Finally, copper delay-line oscillators are also of technical interest and are commercially available for the VHF and UHF bands.

We will analyze the phase-noise mechanism in the delay-line oscillator following the approach of Chapter 4, which was based on the elementary theory of linear feedback systems.

5.1 Basic delay-line oscillator

The delay-line oscillator can be represented as the feedback model of Fig. 5.1. This is similar to Fig. 4.6(a) but the resonator is replaced by a delay line of delay τ_d. The signal $V_i(s)$ allows initial conditions and noise to be introduced into the system, as well as the driving signal if the oscillator is injection-locked. The noise transfer function, again defined as

$$H(s) = \frac{V_o(s)}{V_i(s)},$$ (5.1)

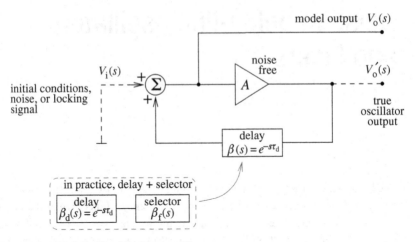

Figure 5.1 Model of the delay-line oscillator.

is the same as that given in (4.43):

$$H(s) = \frac{1}{1 - A\beta(s)} \, . \tag{5.2}$$

A delay τ_d in the time domain maps into multiplication by $e^{-s\tau_d}$ in the Laplace transform. Thus, in this case, it holds that

$$H(s) = \frac{1}{1 - Ae^{-s\tau_d}} \, . \tag{5.3}$$

The poles of $H(s)$ are the roots of the denominator, that is, the solutions of $1 - Ae^{-s\tau_d} = 0$ with $s = \sigma + j\omega$. These poles, which we denote by s_l, are shown in Fig. 5.2. They form an infinite array on a line parallel to the imaginary axis:

$$s_l = \frac{1}{\tau_d} \ln A + j \frac{2\pi}{\tau_d} l, \qquad \text{integer } l \in (-\infty, \infty) \, . \tag{5.4}$$

The poles are in the left-hand half-plane for $A < 1$, move rightwards as A increases, lie on the imaginary axis for $A = 1$, and reach the right-hand half-plane for $A > 1$. The Barkhausen condition for stationary oscillation is therefore met for $A = 1$, as expected. In this condition, the delay-line oscillator can oscillate at any frequency $\omega_l = (2\pi/\tau_d)l$.

The root diagram of Fig. 5.2 with $A = 1$ can be regarded as an infinite array of loss-free resonators, each of which results from a pair of imaginary conjugate poles. When excited by appropriate initial conditions, each resonator sustains a pure sinusoidal oscillation of frequency $\omega_l = (2\pi/\tau_d)l$, as well as dc. At this level of abstraction, there is no reason to restrict the number of resonators excited simultaneously. Thus, the oscillation is a linear superposition of dc and sinusoids:

$$v(t) = \frac{1}{2} a_0 + \sum_{l=1}^{\infty} \left[a_l \cos\left(\frac{2\pi l}{\tau_d} t\right) + b_l \sin\left(\frac{2\pi l}{\tau_d} t\right) \right] \, . \tag{5.5}$$

This is the Fourier series expansion of an arbitrary periodic waveform. The property of completeness of the Fourier series tells us that the delay-line oscillator can sustain any periodic waveform.

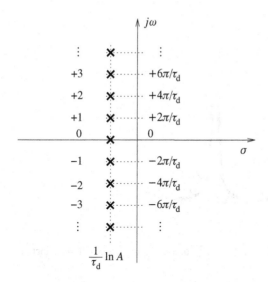

Figure 5.2 Poles of the noise transfer function $H(s)$. The l values, (5.4), are given to the left of the pole positions.

Interestingly, the derivative

$$\frac{ds_l}{dA} = \frac{1}{\tau_d A} \tag{5.6}$$

indicates that a gain change δA causes the poles to move by $\delta s_l = \delta A/(\tau_d A)$. Thus, the poles are shifted horizontally by $\delta \sigma = \delta A/(\tau_d A)$ if $\delta A/A$ is real. From this standpoint, a fluctuation in the amplifier gain with no associated phase fluctuation does not impact on the oscillator phase noise. Obviously, the amplitude must be stabilized by a suitable mechanism that does not appear here.

The power and noise are governed by

$$|H(j\omega)|^2 = \frac{1}{(1 + A^2) - 2A \cos \omega\tau_d} . \tag{5.7}$$

Using $1 + A^2 = (1 - A)^2 + 2A$ and $\cos x = 1 - 2 \sin^2(x/2)$, the above expression becomes

$$|H(j\omega)|^2 = \frac{1}{(1 - A)^2 + 4A \cos(\omega\tau_d/2)} \qquad \text{(Fig. 5.3)}, \tag{5.8}$$

which is formally similar to the Airy function $\mathcal{A}(\theta) = 1/(1 + a \sin \theta)$.

For $A = 1$ the function $|H(j\omega)|^2$ has a series of singularities (infinities) at $\omega = 2\pi l/\tau_d$, integer l, which become sharp resonances for $A < 1$. Therefore, a sub-threshold delay-line oscillator can be used as a filter.

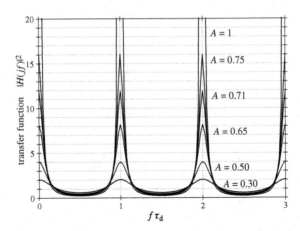

Figure 5.3 Noise transfer function $|H(jf\tau_d)|^2$, for a delay-line loop with no selection filter.

5.2   Optical resonators

5.2.1   Fabry–Pérot interferometer

The Fabry–Pérot (FP) interferometer, also referred to as the FP etalon, is, for reasons not elucidated here, a popular and well-studied optical resonator. The details are available in numerous references, among which the author prefers [94, 90, 10]. Here, we focus on the analogy with the delay-line oscillator. With reference to Fig. 5.4, we will study this resonator using the Laplace transform of the electric field $\mathcal{E}(s)$.

Let us first recall the behavior of a partially transparent mirror. The field $\mathcal{E}(s)$ gives rise to a transmitted field $jT\mathcal{E}(s)$ and a reflected field $R\mathcal{E}(s)$. The coefficients R and T are nonnegative numbers. By energy conservation, it holds that $T^2 + R^2 = 1$ and thus $0 \leq T \leq 1$ and $0 \leq R \leq 1$. The coefficient 'j' in the transmitted field, corresponding to a 90° shift, is a consequence of the phase-matching condition at the mirror interface. It is not relevant for us, since we are interested only in the beam's intensity. An efficient FP etalon requires that R be close to 1, and consequently that T be close to 0, for interference to take place on a large number of round trips, and ultimately for the etalon to be highly selective in frequency.

Let $T_1 T_2 = T^2$ and $R_1 R_2 = R^2$. We first notice that a round trip introduces a factor $R^2 e^{-s\tau_d}$ in the electric field. The transfer function is found imagining the circuit to be broken at a reference plane and matching the electric field on the two sides:

$$jT_1\mathcal{E}_i(s) + \mathcal{E}_r(s)R^2 e^{-s\tau_d} = \mathcal{E}_r(s) \,, \tag{5.9}$$

from which we obtain

$$\frac{\mathcal{E}_r(s)}{\mathcal{E}_i(s)} = \frac{jT_1}{1 - R^2 e^{-s\tau_d}} \,. \tag{5.10}$$

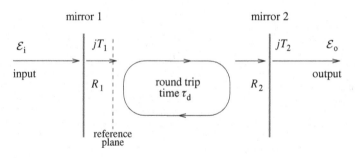

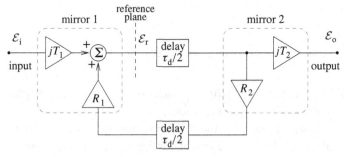

Figure 5.4 Fabry–Pérot etalon.

Finally, $\beta(s) = \mathcal{E}_o(s)/\mathcal{E}_i(s)$ is given by

$$\beta(s) = \frac{-T_1^2}{1 - R^2 e^{-s\tau_d}} \, . \tag{5.11}$$

Dropping the trivial factor $-T^2 e^{-s\tau_d/2}$, and matching the double reflection R^2 to the gain A, the above is formally identical to (5.3), since it has the same properties.

The derivation of (5.11) is based on noticing that the output field is the sum of the field going straight from the input to the output and the fields after $1, 2, 3, \ldots, \infty$ round trips. Thus

$$\mathcal{E}_o(s) = -T^2 e^{-s\tau_d/2} \left[1 + R^2 e^{-s\tau_d} + (R^2 e^{-s\tau_d})^2 + \cdots \right] \mathcal{E}_i(s), \tag{5.12}$$

from which we find (5.11) using the property $\sum_{i=0}^{\infty} x^i = 1/(1 - x)$ for $|x| < 1$.

5.2.2 Whispering-gallery resonator

In this type of resonator, the field is confined in the equatorial region inside a circular dielectric cavity by total reflection. Whispering-gallery resonance derives from the in-phase interference of the traveling field with itself after a round trip. For the energy to be confined, the refraction index must be higher than that of the free space around the resonator. Although a spherical resonator is the most usual in the traditional literature, thanks to the energy confinement in the equatorial region the resonator can take various shapes, e.g. a ring bump on a cylindrical surface or a curved-edge thin disk. Among known electromagnetic resonators, the optical whispering-gallery resonator exhibits the highest quality factors. The incredible result $Q = 6 \times 10^{10}$ at $\lambda = 1.55$ μm, thus

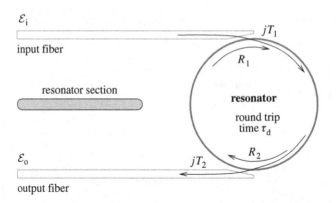

Figure 5.5 Disk resonator, coupled with prism-shaped optical fibers.

$\nu_0 Q \approx 10^{25}$, has been reported [40]. The reader should refer to [68, 52] for the physics and the applications of this type of resonator.

Figure 5.5 shows an example of a whispering-gallery resonator, coupled by prism-shaped optical fibers [53]. By inspection of this figure, the matching condition is given by (5.9) since (5.11) holds in this case also. Radiative and other losses [38] can be included in the term R^2. This is formally sufficient to describe the resonance, even though the physics is quite different.

5.3   Mode selection

A pure delay-line oscillator, with an infinite array of poles on the imaginary axis, represents a mathematical exercise with scarce practical usefulness. In fact a selector filter is necessary, which ensures that the Barkhausen condition $A\beta(j\omega) = 1$ is met at a privileged frequency. Such a filter ensures that a pair of complex conjugate poles, indicated by subscripts m and $-m$, lie on the imaginary axis and that all the other poles s_l, $l \neq \pm m$, are located in the left-hand half-plane, $\Re\{s_l\} < 1$. Introducing such a selector, we replace the feedback function $\beta(s)$ of (5.2) as follows:

$$\beta(s) \rightarrow \beta_\mathrm{d}(s)\beta_\mathrm{f}(s) \tag{5.13}$$

with

$$\beta_\mathrm{d}(s) = e^{-s\tau_\mathrm{d}} \qquad \text{(delay line)}, \tag{5.14}$$
$$\beta_\mathrm{f}(s) = \rho(s)e^{j\theta(s)} \qquad \text{(selection filter)}. \tag{5.15}$$

The function $\beta_\mathrm{f}(s)$ has the properties (4.33)–(4.37) seen at the start of subsection 4.2.1. Introducing the selection filter, the transfer function (5.3) becomes

$$H(s) = \frac{1}{1 - A\rho(s)e^{j\theta(s)}e^{-s\tau_\mathrm{d}}} . \tag{5.16}$$

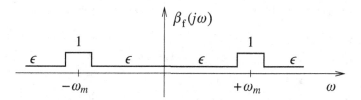

Figure 5.6 Amplitude-only selector filter.

The poles of $H(s)$ are the roots of the denominator

$$\mathcal{D}(s) = 1 - A\rho(s)e^{j\theta(s)}e^{-s\tau_d} . \tag{5.17}$$

Splitting $\mathcal{D}(s) = 0$ into its modulus and argument, we get

$$A\rho e^{-\sigma\tau_d} = 1 ,$$

$$\theta - \omega\tau_d = 0 \quad \text{mod } 2\pi ,$$

and finally

$$\begin{aligned}
\sigma &= \frac{1}{\tau_d} \ln A\rho \\
\omega &= \frac{\theta}{\tau_d} \quad \text{mod } \frac{2\pi}{\tau_d}
\end{aligned} \qquad \text{(poles of } H(s)) . \tag{5.18}$$

It can be seen that the modulus ρ affects only the real part σ of the pole and thus the damping; the argument θ affects only the imaginary part and thus the frequency. The poles of $H(s)$ are located at

$$s_l = \frac{1}{\tau_d} \ln A\rho + j\frac{2\pi}{\tau_d}l + j\frac{\theta}{\tau_d} \qquad \text{(poles of } H(s)) . \tag{5.19}$$

Equation (5.19) will be used to analyze some relevant types of selection filter, either of only theoretical interest or practically realizable.

5.3.1   Amplitude-only filter

This selector filter is defined as (Fig. 5.6)

$$\beta_f(s) = \begin{cases} 1 & \text{for } \omega \approx \pm\omega_m, \text{ thus } l = \pm m , \\ \epsilon, \text{ with } 0 < \epsilon < 1 , & \text{elsewhere } (l \neq \pm m) . \end{cases} \tag{5.20}$$

Such a filter attenuates by a factor $\epsilon < 1$ all signals except those in the vicinity of $\pm\omega_m$. Needless to say, it is an abstraction that cannot be fully implemented by real-world devices.

The poles of $H(s)$ are the roots of its denominator, i.e. they are given by

$$1 - Ae^{-s\tau_d} = 0 \qquad \omega \approx \pm\omega_m \quad (l = \pm m), \tag{5.21}$$

$$1 - \epsilon Ae^{-s\tau_d} = 0 \qquad \text{elsewhere} \quad (l \neq \pm m) . \tag{5.22}$$

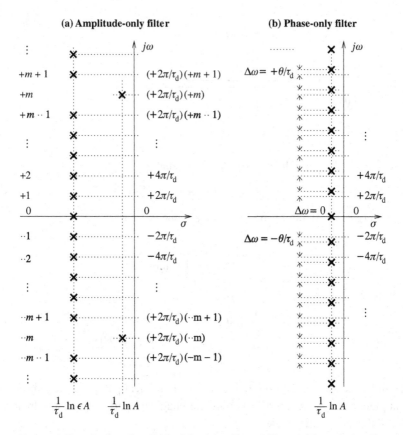

Figure 5.7 Transfer function $H(s)$ of the delay-line oscillator with a selector filter $\beta_f(s)$ inserted in the oscillator loop. (a) Amplitude-only filter. The l values, (5.23), are to the left of the pole positions. (b) Phase-only filter, (5.26). The value of $\Delta\omega$ is $+\theta/\tau_d$ at every pole above the real axis and $-\theta/\tau_d$ at every pole below it.

Such poles, shown in Fig. 5.7(a) are located at

$$
s_l = \begin{cases}
\dfrac{1}{\tau_d}\ln A + j\dfrac{2\pi}{\tau_d}l & \text{for } l = \pm m, \text{ thus } \omega = \pm\omega_m, \\[3mm]
\dfrac{1}{\tau_d}\ln \epsilon A + j\dfrac{2\pi}{\tau_d}l & \text{for } l \neq \pm m, \text{ thus } \omega \neq \pm\omega_m.
\end{cases}
\tag{5.23}
$$

5.3.2   Phase-only filter

The phase-only filter is defined by

$$
\beta_f(s) = e^{j\theta \,\mathrm{sgn}\,\omega}, \tag{5.24}
$$

where θ is a constant. This filter shifts the signals by θ for $\omega > 0$ and by $-\theta$ for $\omega < 0$. The sign function $\mathrm{sgn}\,\omega$ is necessary for the condition $\beta_f(s) = \beta_f^*(s^*)$ to be satisfied. Once again, this filter is an abstraction.

By substituting the above $\beta_f(s)$ into $H(s)$, we get

$$H(s) = \frac{1}{1 - Ae^{j\theta \, \mathrm{sgn}\,\omega}e^{-s\tau_d}} \, . \tag{5.25}$$

The poles of $H(s)$ are the roots of the denominator

$$\mathcal{D}(s) = 1 - Ae^{j\theta \, \mathrm{sgn}\,\omega}e^{-s\tau_d} \, .$$

By splitting the modulus and the argument of $\mathcal{D} = 0$, we find

$$Ae^{-\sigma\tau_d} = 1 \, ,$$

$$\theta \, \mathrm{sgn}\,\omega - \omega\tau_d = 0 \quad \mathrm{mod}\ 2\pi \, .$$

Hence the poles of $H(s)$, shown in Fig. 5.7(b), are located at

$$s_l = \frac{1}{\tau_d}\ln A + j\left(\frac{2\pi}{\tau_d}l + \frac{\theta}{\tau_d}\,\mathrm{sgn}\,l\right), \qquad \text{integer } l \, . \tag{5.26}$$

In conclusion, the phase-only filter affects only the imaginary part of the pole location, not the real part. Thus, the phase θ results in a frequency shift (Fig. 5.7(b)) given by

$$\Delta\omega = \begin{cases} \dfrac{\theta}{\tau_d} & \text{for } \omega > 0 \, , \\[3mm] -\dfrac{\theta}{\tau_d} & \text{for } \omega < 0 \, . \end{cases} \tag{5.27}$$

5.4   The use of a resonator as a selection filter

Let us provisionally assume that the resonator is tuned at the exact frequency $\omega_m = 2\pi m/\tau_d$ of the mode $l = m$ of interest, which is a permitted frequency of the delay-line oscillator. The feedback function is $\beta(s) = \beta_d(s)\beta_f(s)$, with

$$\beta_d(s) = e^{-s\tau_d} \qquad\qquad \text{(delay line)}, \tag{5.28}$$

$$\beta_f(s) = \frac{\omega_m}{Q}\frac{s}{s^2 + (\omega_m/Q)s + \omega_m^2} \qquad \text{(resonator, (4.25) with } \omega_n = \omega_m). \tag{5.29}$$

It is important to understand the roles of the delay line and of the filter. The main technical motivation for the delay-line oscillator is the need for a delay line that exhibits high stability. Hence the oscillation frequency must be determined by the delay line, with at most a minor contribution of the filter's center frequency. For this to be true, it is necessary that

$$\frac{d}{d\omega}\arg\beta_d(j\omega) \gg \frac{d}{d\omega}\arg\beta_f(j\omega) \qquad \text{at } \omega = \omega_m \, , \tag{5.30}$$

because the frequency fluctuations are weighted by the phase slope $(d/d\omega)\arg\beta(j\omega)$ of the feedback elements. This is a consequence of the Barkhausen phase condition $\arg\beta(j\omega) = 0$. Equation (5.30) is equivalent to

$$\tau_d \gg \tau_f \, , \tag{5.31}$$

where τ_f is the filter group delay (in other instances denoted by τ_g),

$$\tau_f = \frac{2Q}{\omega_m} \qquad \text{(filter group delay at } \omega = \omega_m) . \qquad (5.32)$$

Replacing $\beta(s)$ by $\beta_d(s)\beta_f(s)$ in $H(s) = 1/(1 - A\beta(s))$, under the assumption that the amplifier gain is $A = 1$ we find the oscillator transfer function

$$H(s) = \frac{s^2 + \omega_m s/Q + \omega_m^2}{s^2 + \omega_m s/Q + \omega_m^2 - (\omega_m s/Q)e^{-s\tau_d}} . \qquad (5.33)$$

This function has a pair of complex conjugate zeros at

$$s_z, s_z^* \simeq -\frac{\omega_m}{2Q} \pm j\omega_m$$

and a series of complex conjugate poles, to be discussed below.

As a consequence of the condition (5.31), the resonator bandwidth ω_m/Q is large compared with the free spectral range $2\pi/\tau_d$. This means that in frequency ranges $\mathcal{F}$ around ω_m and also around $-\omega_m$ it holds that $|Q\chi| \ll 1$ where χ is the dissonance; see (4.31). This range $\mathcal{F}$ contains a few modes ω_l around ω_m. In $\mathcal{F}$, the resonator function $\beta_f(j\omega)$ is close to 1. This fact has the following consequences:

1. The poles of $H(s)$ are chiefly determined by the oscillation of $\arg \beta_d(s)$. They are close to $j2\pi l/\tau_d$, which is the result already obtained without the selection filter.
2. As an approximation, we can replace $\beta_f(s)$ by $\beta_f(j\omega)$, which can be written in polar coordinates as

$$\beta_f(j\omega) = \rho(\omega)e^{j\theta(\omega)}, \qquad (5.34)$$

with

$$\rho(\omega) = \frac{1}{\sqrt{1 + Q^2\chi^2}} \qquad \text{(modulus)} ,$$

$$\theta(\omega) = -\arctan Q\chi \qquad \text{(phase)} ,$$

$$\chi = \frac{\omega}{\omega_m} - \frac{\omega_m}{\omega} \qquad \text{(dissonance)} .$$

3. In the vicinity of the oscillation frequency, thus for $|(\omega - \omega_m)/\omega_m| \ll 1/(2Q)$ or equivalently for $|(l - m)/m| \ll 1/(2Q)$ we can approximate the dissonance as

$$\chi \simeq 2\frac{\omega - \omega_m}{\omega_m} . \qquad (5.35)$$

This approximation *holds only for positive frequencies* and thus for positive l. The negative-frequency half-plane can be reconstructed using the *symmetry rules*.
4. The modulus $\rho(\omega)$ and $\theta(\omega)$ can be further approximated as

$$\rho(\omega) = 1 - 2Q^2\left(\frac{\omega - \omega_m}{\omega_m}\right)^2 ,$$

$$\theta(\omega) = -2Q\frac{\omega - \omega_m}{\omega_m} .$$

The approximate pole location is found by inserting the above expressions for $\rho(\omega)$ and $\theta(\omega)$, evaluated at $\omega = \omega_l$, into (5.19) with $A = 1$. The poles are then at $s_l = \sigma_l + j\omega_l$, with

$$\sigma_l = \frac{1}{\tau_d} \ln\left[1 - 2Q^2 \left(\frac{\omega_l - \omega_m}{\omega_m} \right)^2 \right] \qquad (\omega > 0) \qquad (5.36)$$

$$\simeq -\frac{2Q^2}{\tau_d} \left(\frac{\omega_l - \omega_m}{\omega_m} \right)^2 , \qquad (5.37)$$

$$\omega_l = \frac{2\pi}{\tau_d} l + \frac{\theta}{\tau_d} \qquad (\omega > 0) \qquad (5.38)$$

$$\simeq \frac{2\pi}{\tau_d} l - \frac{2Q}{\tau_d} \frac{\omega_l - \omega_m}{\omega_m} ; \qquad (5.39)$$

hence, they are at

$$s_l \simeq -\frac{2Q^2}{\tau_d} \left(\frac{\omega_l - \omega_m}{\omega_m} \right)^2 + j\frac{2\pi}{\tau_d} l - j\frac{2Q}{\tau_d} \frac{\omega_l - \omega_m}{\omega_m} \qquad (\omega > 0). \qquad (5.40)$$

Additionally, we might be interested in expressing the poles as a function of the integer offset μ from the main mode, defined as

$$\mu = \begin{cases} l - m & \text{for } \omega > 0, \text{ thus for } l > 0, \\ l + m & \text{for } \omega < 0, \text{ thus for } l < 0. \end{cases} \qquad \text{(definition of } \mu\text{)} \qquad (5.41)$$

The fractional frequency offset is then

$$\frac{\omega_l - \omega_m}{\omega_m} = \frac{\mu}{m} , \qquad (5.42)$$

and hence

$$s_\mu = -\frac{2Q^2}{\tau_d} \left(\frac{\mu}{m} \right)^2 + j\frac{2\pi}{\tau_d} (m + \mu) - j\frac{2Q}{\tau_d} \frac{\mu}{m} \qquad (\omega > 0). \qquad (5.43)$$

Figure 5.8 shows the poles of $H(s)$ located on a parabola centered at ω_m. The pole frequencies are shifted by $-2Q\mu/(\tau_d m)$ with respect to the exact mode frequency $2\pi l/\tau_d$ of the delay line. This frequency shift results from the off-resonance phase of the filter. The negative-frequency part of the complex plane follows by symmetry.

Interestingly, every pair of complex conjugate poles of $H(s)$ can be seen as a separate resonator (Fig. 4.4 and (4.25) with $\omega_n - \omega_l$), whose quality factor Q_l derives from the filter Q enhanced by positive feedback. The pole pair selected by the resonator, $l = m$, lies on the imaginary axis since the equivalent quality factor is $Q_m = \infty$. Using $Q = -\arctan(\Im\{s_l\}/\Re\{s_l\})$, the quality factor associated with the other pole pairs, for which $l \neq m$, is

$$Q_l \simeq \frac{\tau_d \omega_l}{4Q^2} \left(\frac{m}{\mu} \right)^2 \qquad (5.44)$$

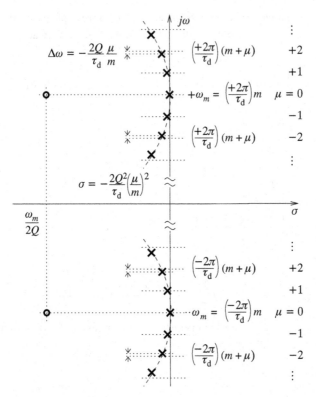

Figure 5.8 Transfer function $H(s)$ of the delay-line oscillator with a resonator as the selector filter. The frequency shift is $\Delta\omega = -2Q\mu/(\tau_d m)$.

or equivalently

$$Q_\mu \simeq \frac{\pi l}{2Q^2}\left(\frac{m}{\mu}\right)^2 = \frac{\pi(\mu+m)}{2Q^2}\left(\frac{m}{\mu}\right)^2 .$$

In the presence of noise, these resonance poles become energized by it. Thus, the radio-frequency spectrum shows a series of sharp lines at frequencies a distance ω_μ from the carrier frequency ω_m, on both sides because μ takes positive and negative values. These lines are easily mistaken for competitor oscillation modes, which under our assumptions they cannot be. As a relevant difference, random excitation causes incoherent oscillation, and such oscillation is damped. Of course, mode competition and multimode oscillation can occur, yet under different assumptions on the gain saturation mechanism; this sort of behavior is commonly found in lasers.

With the operating parameters of actual or conceivable oscillators, the equivalent quality factor Q_l is so high that the noise transfer function $|H(j\omega)|^2$, (5.33), appears to have a series of infinitely narrow frequency impulses $\delta(\omega)$. Examples are provided in Section 5.8.

5.4.1 ★ De-tuned resonator

The assumption that the selection filter is tuned to the exact frequency $\omega_m = 2\pi\, m/\tau_d$ of the mode m of interest is not realistic in practice. Removing this assumption, the filter's natural frequency

$$\omega_n = \omega_m + \omega_e \,, \tag{5.45}$$

is still close to ω_m. In fact, the frequency error ω_e cannot exceed half the distance between modes,

$$-\frac{1}{2}\frac{2\pi}{\tau_d} < \omega_e < \frac{1}{2}\frac{2\pi}{\tau_d} \,, \tag{5.46}$$

otherwise the oscillator switches to the contiguous mode. The condition $\tau_d \gg \tau_f$, (5.31), is still assumed, for the filter function is locally approximated by a linear phase and a quadratic absolute value. Under these assumptions, we can replace the term $(\omega_l - \omega_m)/\omega_m$ of (5.40) with

$$\frac{\omega_l - (\omega_m + \omega_e)}{\omega_m + \omega_e} \simeq \frac{\omega_l - \omega_m}{\omega_m} - \frac{\omega_e}{\omega_m} \,. \tag{5.47}$$

This has the effect of shifting by ω_e the parabola on which the poles are located and of introducing a small frequency offset by the addition of

$$j\frac{2Q}{\tau_d}\frac{\omega_e}{\omega_m}$$

to the imaginary part of s_l. However, this is not a preferred way to pull the oscillator frequency because the detuning may induce mode jump. Instead, in the case of optoelectronic oscillators, it is possible to change the optical delay by using the thermo-refractive effect, acting on the temperature, or by using the change in dispersion caused by a change in the laser wavelength.

5.4.2 Flat-response band-pass filter

In the domain of telecommunications, complex band-pass filters are widely used; these exhibit a nearly flat response over the bandwidth (Fig. 5.9), with only a small ripple. The experimentalist may be inclined to use one of these filters as the mode selector because they are commercially available, and out of the common belief that flatness is

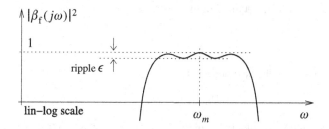

Figure 5.9 Low-ripple band-pass filter used in telecommunications.

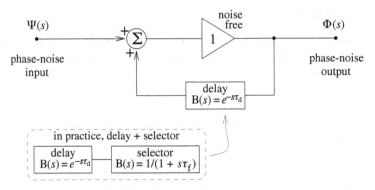

Figure 5.10 Phase-noise model of the delay-line oscillator.

one of the most desirable frequency attributes for a pass band. Unfortunately, a flat-response filter is the *worst choice* for an oscillator. The reason is that the $\mu \neq 0$ poles are closer to the imaginary axis than they are for a resonator. This has the following undesirable consequences:

1. selecting an oscillation frequency is difficult;
2. mode jump may occur;
3. the close-in noise peaks are maximally high.

5.5   Phase-noise response

Thanks to a gain control mechanism and to the appropriate initial conditions, a delay-time oscillator oscillates steadily at the frequency $\omega_m = 2\pi m / \tau_d$. We are now able to analyze the phase-noise transfer function, already defined in Chapter 4 as $H(s) = \Phi(s)/\Psi(s)$, (4.91). The phase-noise model of such an oscillator is shown in Fig. 5.10. In this circuit, all the signals are Laplace transforms of the phase fluctuations present in the oscillator. The input signal $\Psi(s)$ is the phase noise of the oscillator loop components, referred to the amplifier input. Hence the amplifier is rendered free from noise. The amplifier gain is equal to unity because the amplifier just "copies" the input phase to the output. Thus it holds that

$$H(s) = \frac{1}{1 - B(s)} \, , \tag{5.48}$$

which is the same as (4.93), obtained in the same way.

5.5.1   No selection filter

In this case, the phase feedback function is

$$B(s) = e^{-s\tau_d} \, , \tag{5.49}$$

which represents a delay line independent of the carrier frequency. A proof of (5.49) can be obtained with the the phase-step method introduced in Section 4.4. In this case, the

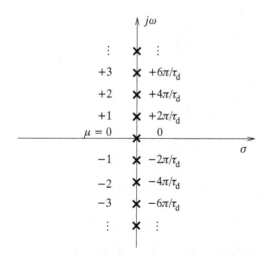

Figure 5.11 Phase-noise transfer function H(s) on the complex plane. Here $l = m$.

phase step propagates unaffected through the delay line, since it appears at the line output after a delay τ_d. Otherwise, we observe that the delay line would have infinite bandwidth, for the group delay is the same at all frequencies and equal to τ_d. Hence the phase, like any information-related parameter of the signal, takes a time τ_d to propagate.

By substituting $B(s) = e^{-s\tau_\mathrm{d}}$ in $H(s) = 1/(1 - B(s))$, we get

$$H(s) = \frac{1}{1 - e^{-s\tau_\mathrm{d}}} \qquad \text{(delay-line oscillator)}. \qquad (5.50)$$

The function $H(s)$ has an infinite number of poles, which are the roots of the denominator. Recalling from (5.41) that the integer mode offset is given by (5.41),

$$\mu = \begin{cases} l - m & \text{for } \omega > 0, \text{ thus for } l > 0, \\ l + m & \text{for } \omega < 0, \text{ thus for } l < 0, \end{cases}$$

we find that the poles (Fig. 5.11) are located at

$$s_\mu = j\frac{2\pi}{\tau_\mathrm{d}}\mu \qquad \text{for integer } \mu. \qquad (5.51)$$

In the spectral domain, the square modulus of the phase-noise transfer function is

$$|H(j\omega)|^2 = \frac{1}{2\,(1 - \cos \omega\tau_\mathrm{d})} \qquad \text{(Fig. 5.12)}. \qquad (5.52)$$

Interestingly, $|H(j\omega)|^2$ has a series of minima $|H(j\omega)|^2_\mathrm{min} = \frac{1}{4}$ at $\omega = \frac{1}{2}\,2\pi/\tau_\mathrm{d}$, $\omega = \frac{3}{2}\,2\pi/\tau_\mathrm{d}$, $\omega = \frac{5}{2}\,2\pi/\tau_\mathrm{d}$, etc., i.e. at odd multiples of half the free spectral range. At these minima, the oscillator phase noise is 6 dB lower than the loop internal noise (thermal, shot, etc.) referred to the carrier power. The physical meaning of these low minima is that the noise is sunk by the neighboring poles.

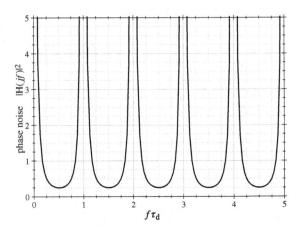

Figure 5.12 Phase-noise transfer function $|H(jf\tau_\mathrm{d})|^2$ for a delay-line oscillator with no selection filter.

5.5.2   A resonator as the selection filter

Let the delay-line oscillator oscillate at the frequency $\omega_m = 2\pi m/\tau_\mathrm{d}$, owing to a resonator in the feedback loop tuned to the exact frequency ω_m. We will analyze the phase-noise transfer function $H(s) = \Psi(s)/\Phi(s)$ under this condition. In the loop, it holds that $H(s) = 1/(1 - B(s))$, (4.93). Introducing a resonator as the selector filter, the feedback function is

$$B(s) = B_\mathrm{d}(s)B_\mathrm{f}(s) , \tag{5.53}$$

with

$$B_\mathrm{d}(s) = e^{-s\tau_\mathrm{d}} \qquad \text{(delay)} , \tag{5.54}$$

$$B_\mathrm{f}(s) = \frac{1}{1 + s\tau_\mathrm{f}} \qquad \text{(filter)} . \tag{5.55}$$

Expanding $H(s)$, we find that

$$H(s) = \frac{1 + s\tau_\mathrm{f}}{1 + s\tau_\mathrm{f} - e^{-s\tau_\mathrm{d}}} . \tag{5.56}$$

The function $H(s)$ has a real zero at $s = -1/\tau_\mathrm{f} = -\omega_m/(2Q)$, a pole at $s = 0$, and a series of complex conjugate poles in the left-hand half-plane, discussed below. By definition, the poles are the roots of the denominator $\mathcal{D} = 1 + s\tau_\mathrm{f} - e^{-s\tau_\mathrm{d}}$ of (5.56).

It was shown earlier that, for technical reasons, it is necessary that $\tau_\mathrm{d} \gg \tau_\mathrm{f} = 2Q/\omega_m$. As a consequence, the poles are expected to be close to the position already found in the absence of the selector, that is, close to $s = j2\pi\mu/\tau_\mathrm{d}$. In other words, the selector has only a small effect on $B(s)$ in a region around the origin that contains just a few pole pairs. This is exactly the same situation that we found when we were searching for the poles of $H(s)$ transposed from $\pm\omega_m$ to the origin.

We first focus on the low-pass phase filter $B_f(s)$. Close to the imaginary axis, we can approximate $B_f(s)$ as $B_f(j\omega)$, which is written in polar form as

$$B_f(s) = \rho(\omega)e^{j\theta(\omega)} \tag{5.57}$$

with

$$\rho(\omega) = \sqrt{\frac{1}{1 + \omega^2 \tau_f^2}} , \tag{5.58}$$

$$\theta(\omega) = -\arctan \omega \tau_f . \tag{5.59}$$

Using $\tau_f = 2Q/\omega_m$ and $\omega \tau_f \ll 1$, we make the approximations

$$\rho(\omega) \simeq 1 - 2Q^2 \left(\frac{\omega}{\omega_m}\right)^2 , \tag{5.60}$$

$$\theta(\omega) \simeq -2Q \frac{\omega}{\omega_m} . \tag{5.61}$$

As the equation $\mathcal{D} = 0$ is equivalent to $1 - B(s) = 0$, the poles of $H(s)$ are the solutions of

$$\rho(\omega)e^{j\theta(\omega)}e^{-\sigma \tau_d}e^{-j\omega \tau_d} = 1 , \tag{5.62}$$

which splits into modulus and argument:

$$\rho(\omega)e^{-\sigma \tau_d} = 1 \qquad \text{(modulus)}$$
$$\theta(\omega) - \omega \tau_d = 0 \mod 2\pi \qquad \text{(argument)} .$$

The modulus condition yields

$$e^{-\sigma \tau_d}\left(1 - 2Q^2 \frac{\omega^2}{\omega_m^2}\right) = 1 ,$$

$$e^{\sigma \tau_d} = 1 - 2Q^2 \frac{\omega^2}{\omega_m^2} ,$$

$$\sigma \tau_d = \ln\left(1 - 2Q^2 \frac{\omega^2}{\omega_m^2}\right) ,$$

$$\sigma = -\frac{2Q^2}{\tau_d} \frac{\omega^2}{\omega_m^2} \qquad \text{for } 2Q^2 \frac{\omega^2}{\omega_m^2} \ll 1 . \tag{5.63}$$

The phase condition yields

$$\omega \tau_d + 2Q \frac{\omega}{\omega_m} = 0 \mod 2\pi ,$$

$$\omega = \frac{2\pi}{\tau_d}\mu - \frac{2Q}{\tau_d}\frac{\omega}{\omega_m} \qquad \text{for integer } \mu . \tag{5.64}$$

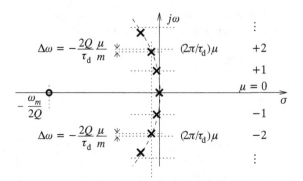

Figure 5.13 Phase-noise transfer function H(s) of the delay-line oscillator with a resonator as the selector filter. Here $\omega_m = (2\pi/\tau_d)m$ and $\sigma = -(2Q^2/\tau_d)(\mu/m)^2$.

Combining (5.63) and (5.64), we find that the poles are located at

$$s_\mu = -\frac{2Q^2}{\tau_d}\frac{\omega_\mu^2}{\omega_m^2} + j\frac{2\pi}{\tau_d}\mu - j\frac{2Q}{\tau_d}\frac{\omega_\mu}{\omega_m}. \qquad (5.65)$$

Additionally, replacing ω_μ/ω_m by μ/m we obtain

$$s_\mu = -\frac{2Q^2}{\tau_d}\left(\frac{\mu}{m}\right)^2 + j\frac{2\pi}{\tau_d}\mu - j\frac{2Q}{\tau_d}\frac{\mu}{m}. \qquad (5.66)$$

Figure 5.13 shows the phase-noise transfer function H(s) on the complex plane. The pole at the origin represents a pure integrator in the time domain and causes the Leeson effect. The other poles are on a horizontal parabola, centered on the origin, at small negative distances $\sigma \propto -\mu^2$ from the imaginary axis. It is easily seen by comparing Fig. 5.13 with Fig. 4.4 that each complex conjugate pair is equivalent to a resonator of quality factor

$$Q_\mu = \frac{\omega_\mu\tau_d}{4Q^2}\left(\frac{m}{\mu}\right)^2, \qquad (5.67)$$

which can be rewritten as

$$Q_\mu = \frac{\pi}{2Q^2}\frac{m^2}{\mu} \qquad (5.68)$$

because we have $\omega_\mu = 2\pi\mu/\tau_d$.

From (5.56), after some lengthy manipulations one obtains

$$|H(j\omega)|^2 = \frac{1 + \omega^2\tau_f^2}{2 - 2\cos\omega\tau_d + \omega^2\tau_f^2 + 2\omega\tau_f\sin\omega\tau_d} \qquad \text{(Fig. 5.14)}. \qquad (5.69)$$

The function $|H(j\omega)|^2$ shows a series of sharp peaks at $\omega = 2\pi\mu/\tau_d$ or $f = \mu/\tau_d$, integer μ. The phase noise taken in is enhanced in the vicinity of these frequencies. The peaks derive from the poles of H(s) close to the imaginary axis yet not on the axis, for they are *not* the signature of competing oscillation modes. The function $|H(j\omega)|^2$ has a series of minima $|H(j\omega)|^2_{\min} = \frac{1}{4}$ at $\omega = \frac{1}{2}2\pi/\tau_d$, $\omega = \frac{3}{2}2\pi/\tau_d$, $\omega = \frac{5}{2}2\pi/\tau_d$, etc.,

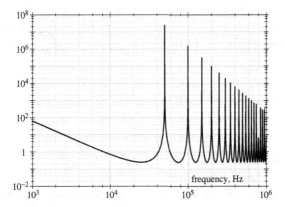

Figure 5.14 Phase transfer function $|H(jf)|^2$ for a delay-line oscillator with selector, evaluated for $\tau_d = 20$ μs, $m = 2 \times 10^5$ (thus $\nu_m = 10$ GHz), and $Q = 2000$ (the data of Example 5.1 in Section 5.8).

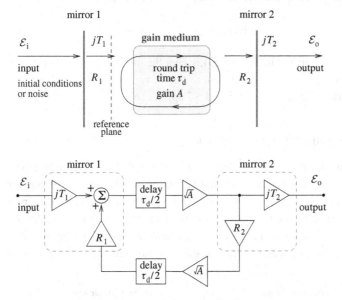

Figure 5.15 A laser can be regarded as a feedback system.

i.e. odd multiples of half the delay. There the oscillator phase noise is 6 dB lower than the loop internal noise (thermal, shot, etc.) referred to the carrier power. At these low minima, the noise is swamped by the neighboring poles.

5.6 Phase noise in lasers

In Fig. 5.15 the optical system of a laser and the equivalent electric circuit are shown. This figure refers to the classical two-mirror laser, that is, a Fabry–Pérot (FP) interferometer with a gain medium inserted between the two mirrors. Gas lasers and semiconductor lasers are of this type. Another structure is common in which the cavity is a ring, for example a whispering-gallery resonator or an optical-fiber loop. The arguments given in subsection 5.2.2 should have made clear that for our purposes the ring structure is fully equivalent to the FP structure.

The transfer function $H(s) = \mathcal{E}_o(s)/\mathcal{E}_i(s)$ is found immediately by comparing Fig. 5.15 (the laser) with Fig. 5.4 (the Fabry–Pérot etalon). Therefore, letting $R_1 R_2 = R^2$,

$$H(s) = \frac{1}{1 - AR^2 e^{-s\tau_d}} . \tag{5.70}$$

The Barkhausen condition is met at the frequencies $\omega_l = 2\pi l/\tau_d$ if $AR^2 = 1$, thus if the gain is sufficient to compensate for the reflection loss at the mirror.

As amplification derives from the stimulated-emission mechanism in pumped atoms, the gain cannot be constant with frequency. In virtually all lasers, it turns out that the bandwidth of the laser gain is greater than the free spectral range of the optical cavity. For our analysis, this will be fixed by taking a constant gain A and pure delay $e^{-s\tau_d}$ and by moving the gain dependence on frequency to a function $\beta_f(s)$:

$$H(s) = \frac{1}{1 - A\beta_f(s)e^{-s\tau_d}} . \tag{5.71}$$

This is the case analyzed in Section 5.4, where the mode-selector filter was introduced.

In conclusion, the laser is equivalent to the delay-line oscillator of Fig. 5.1, and thus the framework of Sections 5.3–5.5 and 5.7 applies to it.

5.6.1   Single-mode and multimode oscillation

In a single-mode laser (Fig. 5.16(a)), when the signal exceeds a threshold the gain drops more or less uniformly, and smoothly, in the pass band. This is found in quantum amplifiers, i.e. masers and lasers. Oscillation sinks atoms from the pumped level and thus the active population is reduced. With these amplifiers the gain condition can be met at only one frequency. Among the modes at which the loop phase is zero, the oscillation frequency is generally that at which the small-signal gain is the highest. In the example

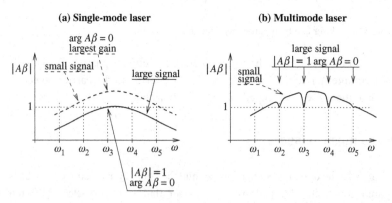

Figure 5.16 Different types of gain saturation in lasers.

of Fig. 5.16(a), $\omega_2, \ldots, \omega_5$ are in the mode competition because they lie in the region where the small-signal loop gain is higher than unity, whereas ω_1 does not. It can be seen that ω_3 wins.

Multimode oscillation requires that the Barkhausen condition $A\beta(j\omega) = 1$ is met at more than one frequency. This is found in some lasers where the gain bandwidth is due to the inhomogeneous broadening mechanism. In this case, the gain is provided by a cluster of energy levels, each of which is narrow enough for the power to sink a population selectively (Fig. 5.16(b)). When oscillation takes place at a given frequency, the active population decreases only near this frequency. Thus oscillation can also arise at the other frequencies where the phase condition is met and the small-signal gain is greater than unity. In the example of Fig. 5.16(b), oscillation takes place at ω_2, ω_3, ω_4, and ω_5, yet not at ω_1 because the small-signal gain is insufficient.

5.7 Close-in noise spectra and Allan variance

From the definition of $H(s)$, the oscillator phase-noise spectrum is

$$S_\varphi(f) = |H(jf)|^2 \, S_\psi(f),\tag{5.72}$$

where $S_\psi(f)$ is the phase noise of the amplifier. A power law is generally not suitable for modeling the phase-noise spectrum of the delay-line oscillator because of the non-polynomial nature of $H(s)$. A power law can only be used at low frequencies, where $|H(jf)|^2$ is approximated as

$$|H(jf)|^2 \simeq \frac{1}{4\pi^2 \tau_d^2} \frac{1}{f^2} \qquad \text{for } f \ll \frac{1}{\tau_d},\tag{5.73}$$

still under the assumption that $\tau_d \gg \tau_f$. Hence, the oscillator phase noise is

$$S_\varphi(f) = \frac{1}{4\pi^2 \tau_d^2} \frac{1}{f^2} S_\psi(f)\tag{5.74}$$

$$= \frac{v_m^2}{4\pi^2 m^2} \frac{1}{f^2} S_\psi(f), \qquad \text{using } v_m = \frac{m}{\tau_d}.\tag{5.75}$$

Introducing the amplifier power-law model $S_\psi(f) = b_0 + b_{-1}/f$, which is limited to white and flicker noise, we get

$$S_\varphi(f) = b_0 \frac{v_m^2}{4\pi^2 m^2} \frac{1}{f^2} + b_{-1} \frac{v_m^2 b_{-1}}{4\pi^2 m^2} \frac{1}{f^3}.\tag{5.76}$$

The fractional-frequency fluctuation spectrum $S_y(f)$ is found by using $S_y(f) = (f^2/v_m^2)S_\varphi(f)$. Thus

$$S_y(f) = b_0 \frac{1}{4\pi^2 m^2} \frac{1}{f^2} + b_{-1} \frac{1}{4\pi^2 m^2} \frac{1}{f^3}.\tag{5.77}$$

After matching (5.77) to the power law $S_y(f) = \sum_i h_i f^i$, we find

$$h_0 = \frac{b_0}{4\pi^2 m^2}\tag{5.78}$$

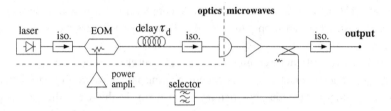

Figure 5.17 Photonic delay-line oscillator.

and

$$h_{-1} = \frac{b_{-1}}{4\pi^2 m^2} .$$ (5.79)

Using Table 1.4 in Section 1.8, the Allan variance is

$$\sigma_y^2(\tau) = (1/\tau^2 \text{ terms}) + \frac{b_0}{4\pi^2 m^2} \frac{1}{2\tau} + \frac{b_{-1}}{4\pi^2 m^2} 2\ln 2 .$$ (5.80)

5.8   Examples

Example 5.1. Photonic delay-line oscillator. We will analyze the photonic delay-line oscillator of Fig. 5.17, using the following parameters and inspired by the references [29] and [109]:

$\tau_d = 20$ µs	4 km optical fiber, refraction index 1.5
$1/\tau_d = 50$ kHz	mode spacing
$\nu_m = 10$ GHz	oscillation frequency $\nu_m = m/\tau_d$
$m = 2 \times 10^5$	mode order
$Q = 2 \times 10^3$	selection filter (tunable microwave cavity)
$b_0 = 3.2 \times 10^{-14}$	white phase noise, -135 rad^2/Hz
$b_{-1} = 10^{-10}$	flicker phase noise, -100 dB rad^2/Hz at $f = 1$ Hz.

The microwave delay is implemented with an optical fiber that carries an intensity-modulated laser beam. The reason for this choice is that the low attenuation of the fiber, of order 0.2 dB/km at 1.55 µm wavelength, makes possible a delay of several tens of microseconds. The attenuation of a microwave cable, with attenuation 0.5–1 dB/m, would limit the delay to a maximum of hundreds of nanoseconds.

Figure 5.14 shows the function $|H(jf)|^2$. A series of spectral lines due to noise is evident at $f = 1/\tau_d = 50$ kHz and multiples thereof. Table 5.1 shows the relevant resonance parameters for neighboring modes $m + \mu$ of the delay line. The reader should observe the high equivalent quality factor Q_μ of the phase-noise response, which explains the sharp resonances of Fig. 5.14.

The amplifier corner frequency, at which $b_{-1}f^{-1} = b_0$, is $f_c = 3.16$ kHz. The estimated phase-noise spectrum is shown in Fig. 5.18. Restricting our attention to the white

Table 5.1 Resonance parameters for photonic oscillator of Example 5.1

μ	$|\nu-\nu_m|$	Q_χ	σ_μ	Q_n	Q_μ
0	0	0	0	∞	∞
± 1	5×10^4	0.02	-10	3.14×10^9	15.7×10^3
± 2	10×10^4	0.04	-40	785×10^6	7.85×10^3
± 3	15×10^4	0.06	-90	349×10^6	5.24×10^3
± 4	20×10^4	0.08	-160	196×10^6	3.93×10^3
± 5	25×10^4	0.10	-250	126×10^6	3.14×10^3

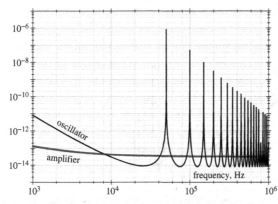

Figure 5.18 Phase-noise spectrum $S_\varphi(f)$ for the 10 GHz photonic delay-line oscillator with selector of Example 5.1. The parameters are $\tau_d = 20$ μs, $m = 2 \times 10^5$, and $Q = 2000$; $b_0 = -135\,\text{rad}^2/\text{Hz}$, $b_{-1} = -100\,\text{db rad}^2/\text{Hz}$.

and the flicker frequency noise, the oscillator phase noise is

$$S_\varphi(f) = \frac{2 \times 10^{-6}}{f^2} + \frac{6.3 \times 10^{-3}}{f^3} .$$

Hence the spectrum of the fractional-frequency fluctuation is

$$S_y(f) = 2 \times 10^{-26} + \frac{6.3 \times 10^{-23}}{f}$$

and the Allan variance is

$$\sigma_y^2(\tau) = \frac{10^{-26}}{\tau} + 8.8 \times 10^{-23} ;$$

thus

$$\sigma_y(\tau) = \frac{10^{-13}}{\sqrt{\tau}} \qquad \text{(white frequency noise)}$$

$$\sigma_y(\tau) = 9.4 \times 10^{-12} \qquad \text{(flicker frequency noise)} .$$

Table 5.2 Resonance parameter's for the SAW oscillator of Example 5.2

μ	$|\nu - \nu_m|$	$Q\chi$	σ_μ	Q_n	Q_μ
0	0	0	0	∞	∞
± 1	2×10^5	0.035	-126	2.24×10^7	4970
± 2	4×10^5	0.071	-506	5.59×10^6	2485
± 3	8×10^5	0.107	-1138	2.49×10^6	1657
± 4	10×10^5	0.142	-2023	1.40×10^6	1243
± 5	12×10^5	0.178	-3160	8.95×10^6	994

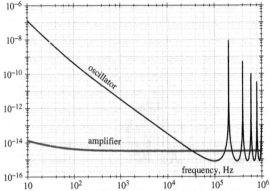

Figure 5.19 Expected phase-noise spectrum $S_\varphi(f)$ for a SAW delay-line oscillator with selector. The parameters are $\tau_d = 5$ μs, $m = 4500$ (thus $\nu_m = 900$ MHz), and $Q = 80$; $b_0 = -145$ rad²/Hz, $b_{-1} = -130$ db rad²/Hz. The data refer to Example 5.2.

Example 5.2. SAW delay-line oscillator. We will analyze a surface acoustic wave (SAW) delay-line oscillator using the following parameters:

$\tau_d = 5$ μs	15 mm SAW, sound speed 3 km/s
$1/\tau_d = 200$ kHz	mode spacing
$\nu_m = 900$ MHz	oscillation frequency; $\nu_m = m/\tau_d$
$m = 4500$	mode order
$Q = 80$	selector filter (LC filter)
$b_0 = 3.2 \times 10^{-15}$	white phase noise, -145 rad²/Hz
$b_{-1} = 10^{-13}$	flicker phase noise, -130 dB rad²/Hz at $f = 1$ Hz.

These values relate to GSM mobile phones, which operate in the 900 MHz band with a channel spacing of 200 kHz. By tuning the selection filter, this oscillator can replace a synthesizer that works at the appropriate frequency, with an appropriate 200 kHz step.

The amplifier corner frequency, at which $b_{-1}f^{-1} = b_0$, is $f_c = 31.6$ Hz. Figure 5.19 shows the expected phase noise $S_\varphi(f)$. The neighboring resonances show up at $f = 1/\tau_d = 200$ kHz and multiples thereof. Table 5.2 shows the relevant resonance parameters.

Focusing our attention on the white and flicker frequency noise, the oscillator phase noise is

$$S_\varphi(f) = \frac{2.5 \times 10^{-6}}{f^2} + \frac{8 \times 10^{-5}}{f^3} \, .$$

The spectrum of the fractional-frequency fluctuation is

$$S_y(f) = 3.96 \times 10^{-24} + \frac{1.25 \times 10^{-22}}{f}$$

and the Allan variance is

$$\sigma_y^2(\tau) = \frac{2 \times 10^{-24}}{\tau} + 1.7 \times 10^{-22} \, ;$$

thus

$$\sigma_y(\tau) = \frac{1.4 \times 10^{-13}}{\sqrt{\tau}} \qquad \text{(white frequency noise)}$$

$$\sigma_y(\tau) = 1.3 \times 10^{-12} \qquad \text{(flicker frequency noise)} \, .$$

6 Oscillator hacking

A more serious title for this chapter could have been *The reverse engineering of phase noise*. When a computer hacker gets into trouble, he or she tackles the root of the problem, by reading the source code, patching the kernel, and recompiling the system rather than just rebooting the machine. There are at least two serious reasons for choosing such a radical approach. The first reason is to acquire a deeper understanding of the system being investigated, i.e., to "solve the puzzle," and the self-confidence that goes along with this. Such an understanding ultimately results in an improvement in the performance and the reliability of the entire system. The second reason is that a radical approach is *ultimately the simplest and perhaps the only legal way* to solve the problem. So one should *solve*, rather than find a provisional remedy. This book is far less ambitious. Nonetheless, having been frustrated for years by the insufficient technical documentation of oscillators, one begins to realize that there is a lot of information written between the lines in invisible ink. Yet it is not that invisible! By hacking the technical information, we can deduce – or at least guess – the relevant parameters, such as P_0, Q, f_L, the amplifier's $1/f$ noise, etc. The need to guess the internal technology is a source of difficulty, inconsistency, and frustration. How to cope with this is part of the message addressed to the reader.

The major problem with reverse engineering is that it is more of an art than a science and relies heavily on a person's experience. Reverse engineering is *not* the kind of thing that can be reduced to a single comprehensive procedure. The examples provided illustrate how the author tackles the problem. Yet, even the *choice* of example reflects an author's experience.

A related problem is that reverse engineering is not free from errors. The reader should be aware that this general statement, unfortunately yet inevitably, also applies to this chapter, for which the author apologizes in advance.

6.1 General guidelines

6.1.1 Inspection on the data sheet

The first step consists of reading carefully the data sheet, focusing on the resonator and on the amplifier technology, and assembling related facts in one's mind. For example, a 5 MHz quartz oscillator may have a quality factor in excess of 10^6, but then it must be driven at very low power, say 10–20 μW, for the best long-term stability. The quality factor of a dielectric resonator can be 1000 or more, depending on its size and frequency. And so on for other resonator types and for the amplifiers. Oscillators similar to those encountered in the past may have a similar spectrum or be surprisingly different.

6.1.2 Parametric estimation of the spectrum

Spectra are in most cases shown as $\mathscr{L}(f)$, which is to be converted into $S_\varphi(f)$ using $\mathscr{L}(f) = \frac{1}{2}S_\varphi(f)$. Then, we have to match the phase-noise spectrum to the polynomial

$$S_\varphi(f) = \sum_{i=-4}^{0} b_i f^i, \tag{6.1}$$

in order to identify the coefficients b_i. A term $b_i f^i$ on a log–log plot appears as a straight line; see the following table.

Noise type	Term		Slope
white phase	b_0	0	0
flicker phase	$b_{-1}f^{-1}$	−1	−10 dB/dec
white frequency	$b_{-2}f^{-2}$	−2	−20 dB/dec
flicker frequency	$b_{-3}f^{-3}$	−3	−30 dB/dec
frequency random walk	$b_{-4}f^{-4}$	−4	−40 dB/dec

Terms with negative slope higher than −4 can be present. The actual spectra are of the form

$$S_\varphi(f) = \sum_{i=-4}^{0} b_i f^i + \sum_j s_j(f), \tag{6.2}$$

where the terms $s_j(f)$ account for mains residuals (50 Hz or 60 Hz and multiples), for bumps due to feedback, and for other stray phenomena. Numerous figures in this chapter provide examples of actual phase-noise spectra. The mathematical process of matching the spectrum to a model is called *parametric estimation* [75, 56]. Some a priori knowledge of the nature of the stray signals may be necessary to match the complete model $\sum_{i=-4}^{0} b_i f^i + \sum_j s_j(f)$ to the observed spectrum. Although (almost) only the power-law coefficients b_i are relevant to our analysis, the terms $s_j(f)$ are essential in that they reduce the bias and the residuals of the estimation.

Whereas computers provide accuracy, a general parametric estimator is not trivial to implement. Conversely, even with little training the human eye is very efficient in extracting the useful information from a spectrum, filtering out the stray signals, and getting a good approximation. *The inspection of a log–log plot by sliding a (preferably transparent) old-fashioned set square against a ruler* (Fig. 6.1) before using a computer is strongly recommended. The author has been seen numerous times in the exhibit areas

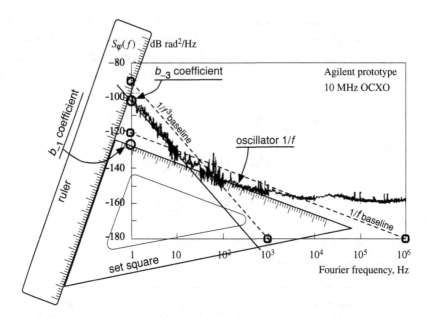

Figure 6.1 Inspection on a log–log plot by sliding a set square and ruler provides a quick and surprisingly accurate parametric estimation of the spectrum. The spectrum is that of Fig. 6.9, used with the permission of the IEEE, 2007.

of international conferences in the process of *sliding credit cards* on a phase-noise spectrum and drawing a straight-line approximation with a sharp pencil. Even this poor replacement for proper drawing tools has proved to be surprisingly useful.

The best way to analyze the spectrum is to work with exact slopes f^0, $1/f$, $1/f^2$, etc. obtained from the coordinate frame using the longest possible baseline. The reference slope is compared with the curve by sliding the set square until its side coincides with the portion of the spectrum having the same slope. Generally, at the corner between two straight lines the true spectrum is found to be 3 dB above the corner point. This is due to one of the following reasons.

1. The difference in slope is 1, i.e., 10 dB/dec, at the corner point. When this occurs, there are two independent random processes whose spectra take the same value $(b_i f^i = b_{i+1} f^{i+1})$ at the corner point.
2. The difference in slope is 2, i.e., 20 dB/dec. When this occurs, at the corner point we have $b_i f^i = b_{i+2} f^{i+2}$. This implies that a single noise process is filtered, owing to the Leeson effect; the factor 2 (3 dB) at the corner point results from a single real zero of the complex transfer function.

In some cases the difference between the spectrum and the straight-line approximation differs from 3 dB at the corner point. Then two estimations are needed, the first based on the straight-line fitting and the second based on the 3 dB difference between the straight lines and the true spectrum at the corner frequencies. The best estimate results from averaging the two worked-out spectra. Physical judgment should be used to assign unequal weights to the two analyses.

6.1.3 Interpretation

This part of the process starts from identification of the spectrum type among those previously analyzed. After that we need to learn about the specific oscillator, which is a unique case. Only a few rules are given here. The reader should refer to the examples given in this chapter then analyze his or her own spectra.

As a general rule, one should analyze the spectrum *from the right-hand side to the left*, thus from white phase noise to frequency flicker or to random walk.

Starting from the white phase noise region we evaluate the power P_0 at the input of the sustaining amplifier, using $S_\varphi(f) = b_0 = FkT_0/P_0$, (2.30). Thus

$$P_0 = \frac{FkT_0}{b_0} . \tag{6.3}$$

One may admit a noise figure $F = 1.25$ (1 dB) for conventional amplifiers, and $F = 3.2$ (5 dB) for noise-corrected amplifiers owing to the input power splitter. Thanks to the gain of the sustaining amplifier, the white noise of the output buffer can generally be neglected.

The next step is to evaluate f_c (the flicker frequency of the sustaining amplifier) and f_L, in order of occurrence from right to left. It is then necessary to guess the oscillator sub-type (Figs. 3.12 and 3.13). A major difficulty is that of understanding whether the oscillator stability derives from the Leeson effect or from resonator fluctuation. Inverting (3.21), the Leeson frequency gives a quality factor

$$Q = \frac{\nu_0}{2f_L} . \tag{6.4}$$

The corner frequency f_c reveals the phase flickering of the amplifier:

$$(b_{-1})_{\text{ampli}} = b_0 f_c . \tag{6.5}$$

In some weird cases the $1/f$ noise of the output buffer shows up. This has been observed in ultra-stable 5 MHz quartz oscillators and in noise-degeneration microwave oscillators.

Allan variance

The Allan variance of the fractional frequency fluctuation (Table 1.4), i.e., the oscillator stability, is related to the phase noise by

$$\sigma_y^2(\tau) = \cdots + \frac{1}{2} \frac{b_{-2}}{\nu_0^2} \frac{1}{\tau} + 2\ln 2 \frac{b_{-3}}{\nu_0^2} + \frac{4\pi^2}{6} \frac{b_{-4}}{\nu_0^2} \tau + \cdots \tag{6.6}$$

Comparing the Allan variance calculated in this way with the Allan variance reported on the data sheet or with the measured Allan variance is often a source of frustration, for a number of reasons. First, the variance is always approximate because the integral (1.81) (or (1.89) for the modified Allan variance) may not converge without the introduction of a low-pass filter. Of course, the bandwidth and roll-off law of this filter will affect the result. Second, and more generally, the measurement results are sensitive to the experimental method and to the operating parameter. Finally, time-domain measurements and spectrum measurements are considered as two separate worlds. This occurs because the two domains tend not to involve the same people and because very few oscillators are measured carefully in both domains, overlapping the time frame. Of course, having consistent results involves an effort that is seldom made. In conclusion, the reader should not be discouraged by any lack of consistency between spectra and variances.

6.2   About the examples of phase-noise spectra

The phase-noise spectra used in this chapter, with very few exceptions, were taken from a manufacturer's website or data sheet, or from journal articles. When it was necessary to redraw a spectrum, the original was imitated as far as possible. This choice has the advantage of training the reader to the real world, at the inevitable cost of making direct comparisons that are sometimes uncomfortable.

All the formulae of this book refer to $S_\varphi(f)$, while the manufacturers prefer $\mathscr{L}(f)$. Therefore, the vertical scales of most spectra have been overwritten in order to facilitate the reader. Recalling that by definition $\mathscr{L}(f) = \frac{1}{2} S_\varphi(f)$, the conversion from the technical unit dBc/Hz to the SI unit dB rad^2/Hz requires the addition of 3 dB. This explains why the numbers on the vertical scale end in "7."

All the figures carry the author's deductions. Most data fitting was done graphically, as in Fig. 6.1, working on an enlarged copy of the original plot and interpolating sharp thin lines with patience. The computer version of the figure was prepared only at the end of the process. In at least one case (Fig. 6.16), however, the spectrum was worked out directly on a computer, yet enlarging the picture to full size on a 23 inch screen.

The interpretations proposed in this chapter are fully independent of the manufacturer's feedback, when provided. When a piece of information comes from the manufacturer, which occurs in a very few cases, this is clearly stated.

6.3   Understanding the quartz oscillator

The piezoelectric quartz is certainly the oscillator type that has the largest production, the largest variety of applications, and the highest economic impact; thus it takes the first place in our analysis. We will focus only on ultra-stable quartz oscillators. This choice is like testing the technology of future production cars on formula-one racing models. Most of what we can learn is found there.

A common signature is found in a number of phase-noise spectra and this deserves to be introduced before the case studies [83]. The procedure detailed in this section should help the reader to understand the interplay between a phase-noise spectrum and the oscillator parameters; it leads to the conclusion that the amplifier's phase noise transformed into frequency noise via the Leeson effect is less than the fluctuation in the resonator's natural frequency.

(a)

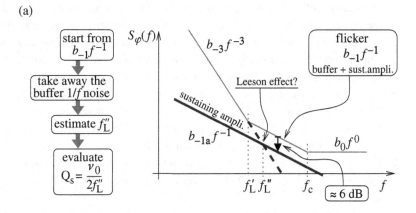

(b)

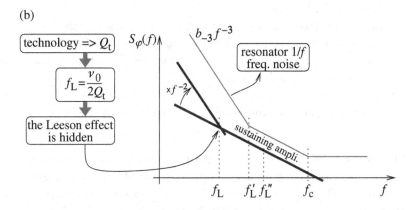

Figure 6.2 Interpretation of the phase noise in ultra-stable quartz oscillators. Adapted from [83] and used with the permission of the IEEE, 2007.

In ultra-stable 5–10 MHz oscillators, phase flickering (b_{-1}/f) is always present in the phase-noise spectrum. This fact is discussed below, with the help of Fig. 6.2.

1. The phase-noise plot changes slope by 2 (from $1/f^3$ to $1/f$) at $f = f'_L$. We want to know whether this corner point is the Leeson frequency or something else.
2. Owing to the high Q value and the low carrier frequency ν_0, and thus the low value of f_L, the oscillator loop is expected to produce an observable $1/f$ phase noise.
3. The oscillator loop requires high isolation from the output to prevent the load from affecting the oscillation frequency, so a buffer is necessary. Design practice suggests that there can be three buffer stages.
4. The $1/f$ phase noise of cascaded amplifiers adds up directly; the contributions *are not divided* by the gain of the preceding stages. This is explained in detail in subsection 2.3.2. Thus, the total flicker b_{-1}/f is the sum of the sustaining-amplifier noise plus the buffer noise.

5. With three buffers whose $1/f$ noise is similar to that of the sustaining amplifier, the noise of the sustaining amplifier is about one-quarter of the total flicker, that is, 6 dB lower. The difference could be even greater than 6 dB, because the output amplifier can be more complex and because superior technology would be used in the sustaining amplifier rather than in the buffers.

6. The Leeson effect takes place inside the oscillator loop, before buffering. After taking away the buffer noise we observe that the sustaining-amplifier $1/f$ noise crosses the $1/f^3$ noise at $f = f_L''$. Once again, we want to know whether this corner point is the Leeson frequency or something else.

7. If f_L'' is indeed the Leeson frequency then we are able to calculate the quality factor using $Q = v_0/(2f_L)$. We denote the result by Q_s, where the subscript "s" stands for *spectrum*.

8. Leaving the noise spectrum aside for a while, we estimate the quality factor on the grounds of the technology involved. We denote this value by Q_t, where the subscript "t" stands for *technology*.

9. In all cases analyzed, it turns out that $Q_t \ll Q_s$ and that the ratio Q_t/Q_s is far too high for Q_s to be the resonator's quality factor, even accounting for the large uncertainty in guessing Q_t. Thus, we have to trust the technology and admit that the Leeson frequency is $f_L = v_0/(2Q_t) < f_L''$.

10. The final conclusion is that resonator instability overrides the Leeson effect. This conclusion, of course, comes only after a thorough analysis of the Leeson effect.

The process described relies on the ability to estimate the resonator's quality factor. Experience indicates that the product $v_0 Q$ is a technical constant for piezoelectric quartz resonators, in the range from 1×10^{13} to 2×10^{13}. As a matter of fact, the highest values are found in 5 MHz resonators. When the resonator is loaded by the rest of the circuit, its Q value is reduced by a factor 0.5–0.8. The actual value depends essentially on the sustaining amplifier and, in turn, on the frequency and on the available technology. A lot of data are available from [36, 62, 104], and from our early attempts to measure the resonator frequency stability [84]. The oscillators we have selected for analysis exhibit the highest available stability and are primarily intended for high-end scientific applications. This background encourages one to trust the published data.

6.4   Quartz oscillators

Oscilloquartz OCXO 8600 (5 MHz AT-cut BVA)

Figure 6.3 shows the phase-noise spectrum of the oven-controlled quartz oscillator Oscilloquartz OCXO 8600. This oscillator has been chosen as an example because of its outstanding stability in the 0.1–10 s region. The spectrum is of type 2 ($f_L < f_c$), as is typical of low-frequency oscillators with high-Q resonators. The plot is fitted by the

$S_\varphi(f)$, dB rad^2/Hz

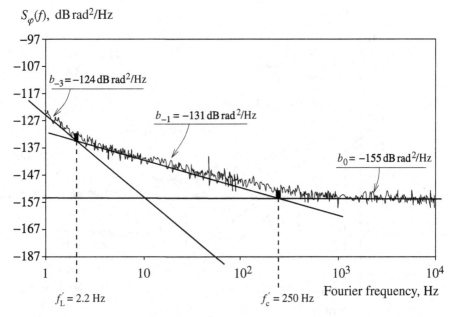

Figure 6.3 Phase noise of the oven-controlled 5 MHz quartz oscillator Oscilloquartz OCXO 8600, with a preliminary (wrong) interpretation. Courtesy of Oscilloquartz SA, 2007. The interpretation, comments, and mistakes are those of the present author.

polynomial $\sum_{i=-3}^{0} b_i f^i$, with

b_0	-155 dB	3.2×10^{-16}	rad^2/Hz
b_{-1}	-131 dB	7.9×10^{-14}	rad^2/Hz
b_{-2}		(not visible)	
b_{-3}	-124 dB	4×10^{-13}	rad^2/Hz

A preliminary interpretation of the spectrum is the following.

1. The white phase noise is $b_0 = 3.2 \times 10^{-16}$ rad^2/Hz (-155 dB rad^2/Hz). Thus

$$P_0 = \frac{FkT}{b_0} \simeq 16 \ \mu W \quad (-18 \text{ dB m}),$$

assuming that $F = 1$ dB. This power level is consistent with one's general experience of 5–10 MHz oscillators, in which the power is kept low for best stability.

2. The value $f_L' \simeq 2.2$ Hz, if it corresponds to the Leeson frequency, gives a quality factor

$$Q_s' = \frac{\nu_0}{2 f_L'} \simeq 1.1 \times 10^6 \ .$$

The question is whether f_L' is in fact the Leeson frequency.

Item 1 above is correct because the effect of the output buffer on the white noise is divided by the gain of the sustaining amplifier, thus we believe that it is negligible. The white noise originates in the series resistance of the resonator and in the sustaining

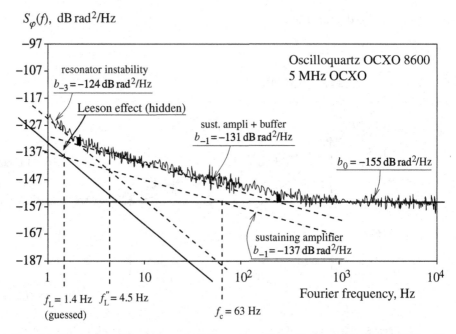

$S_\varphi(f)$, dB rad²/Hz

Oscilloquartz OCXO 8600
5 MHz OCXO

resonator instability
$b_{-3} = -124$ dB rad²/Hz

Leeson effect (hidden)

sust. ampli + buffer
$b_{-1} = -131$ dB rad²/Hz

$b_0 = -155$ dB rad²/Hz

sustaining amplifier
$b_{-1} = -137$ dB rad²/Hz

$f_L = 1.4$ Hz $f_L'' = 4.5$ Hz
(guessed)

$f_c = 63$ Hz

Fourier frequency, Hz

Figure 6.4 Phase noise of the oven-controlled quartz oscillator Oscilloquartz OCXO 8600, with revised interpretation. Courtesy of Oscilloquartz SA, 2007. The interpretation, comments and any mistakes are those of the present author.

amplifier input. Conversely, the flicker noise of the output buffer adds according to the rules given in Section 2.4; see (2.51). Thus, we change the interpretation as follows. The spectrum with the author's comments is shown in Fig. 6.4.

1. Let us assume that the $1/f$ noise of the sustaining amplifier is one-quarter of the total $1/f$ noise, guessing that there are three stages between the sustaining amplifier and the output and that the $1/f$ noise of each stage is at least equal to that of the sustaining amplifier.
2. Taking away 6 dB, the new estimate of the sustaining-amplifier flicker is

$$(b_{-1})_{\text{ampli}} = 2 \times 10^{-14} \qquad (-137 \text{ dB rad}^2/\text{Hz}).$$

Accordingly, the new estimated values are $f_c \simeq 63$ Hz for the corner frequency of the sustaining amplifier and

$$f_L'' \simeq 4.5 \text{ Hz}$$

for the Leeson frequency. The latter is still provisional.

3. Using $f_L'' = 4.5$ Hz in $f_L = \nu_0/(2Q)$, we get $Q_s \simeq 5.6 \times 10^5$.
4. $Q_s \simeq 5.6 \times 10^5$ is far too low a value for a top-technology 5 MHz oscillator. The literature suggests that the product $\nu_0 Q$ is between 10^{13} and 2×10^{13}.

5. There is no doubt that $Q > 5.6 \times 10^5$. This means that the Leeson effect is hidden and that we are not able to calculate Q. Henceforth, our knowledge about Q relies only upon experience and on the literature.

6. Let us guess that $Q = 2.5 \times 10^6$, reduced to $Q = 1.8 \times 10^6$ (loaded) by the dissipative effect of the amplifier input. It follows that

$$f_L = \frac{\nu_0}{2Q} \approx 1.4 \text{ Hz} .$$

This indicates that the oscillator frequency flicker $b_{-3} \simeq 3.2 \times 10^{13}$ is due to frequency flickering in the resonator. The Leeson effect is hidden by the resonator instability and by the phase noise of the output buffer. Thus the corner frequency $f_L'' \simeq 4.5$ Hz of Fig. 6.4 is not the Leeson frequency.

7. The flicker noise of the sustaining amplifier combined with the (guessed) Leeson frequency gives the stability limit of the electronics of the oscillator. This is the solid line f^{-3} in Fig. 6.4, some 10 dB below the phase noise.

8. The frequency flicker expressed as the Allan variance (Table 1.4) is

$$\sigma_y^2(\tau) = 2 \ln 2 \, h_{-1} = 2 \ln 2 \, \frac{b_{-3}}{\nu_0^2} = 1.39 \times \frac{4 \times 10^{-13}}{(5 \times 10^6)^2} ;$$

thus

$$\sigma_y^2(\tau) \simeq 2.2 \times 10^{-26} \quad \text{and so} \quad \sigma_y(\tau) \simeq 1.5 \times 10^{-13} .$$

This is lower than the value $\sigma_y(\tau) < 3 \times 10^{-13}$ for $0.2 \text{ s} \leq \tau \leq 30 \text{ s}$ given in the specifications. As a general rule, the actual Allan variance is higher than the value calculated in this way because it results from an integral that takes in all the noise phenomena, not just h_{-1}/f. Yet, this incompleteness of the integral is not sufficient to explain the discrepancy. Of course, we should also bear in mind that specifications are intended to be conservative; a sample can be better than the specifications.

Oscilloquartz OCXO 8607 (5 MHz SC-cut BVA)

This oscillator is similar to the 8600 considered above, yet the short-term stability is further improved. Figure 6.5 shows the phase-noise spectrum, together with the specifications (bold upright crosses). The spectrum is again of type 2 ($f_L < f_c$), typical of low-frequency oscillators with high-Q resonators. The f^{-3} noise is hardly visible on the left-hand side of the $S_\varphi(f)$ plot, which starts from 1 Hz. Thus curve

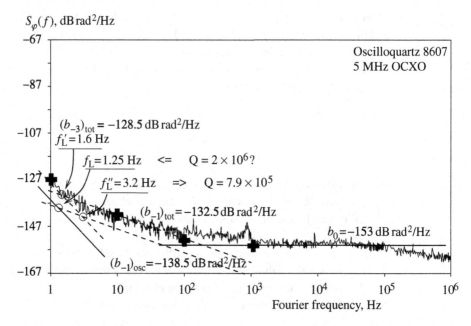

Figure 6.5 Phase noise of the oven-controlled quartz oscillator Oscilloquartz OCXO 8607. Courtesy of Oscilloquartz SA, 2007. The interpretation, comments, and any mistakes are those of the present author.

fitting is impractical, and we have to rely on the specification table, which is given below.

f, Hz	S_φ, dB rad^2/Hz
1	-127
10	-142
10^2	-150
10^3	-153

Looking at the spectrum and at the Allan variance, it is clear that at $f = 1$ Hz and $f = 10$ Hz the terms $b_{-3}f^{-3}$ and $b_{-1}f^{-1}$ determine $S_\varphi(f)$, with at most a minor contribution of b_0. It is also clear that $S_\varphi(f)|_{1\mathrm{kHz}} \simeq b_0$ and that there is no visible $b_{-2}f^{-2}$. Thus b_{-3} and b_{-1} are obtained by solving

$$\left(b_{-3}f^{-3} + b_{-1}f^{-1}\right)_{1\mathrm{Hz}} = 10^{-12.7} \qquad \text{(1 Hz spec.)},$$

$$\left(b_{-3}f^{-3} + b_{-1}f^{-1} + b_0\right)_{10\mathrm{Hz}} = 10^{-14.2} \qquad \text{(10 Hz spec.)},$$

$$b_0 = 10^{-15.3} \qquad \text{(1 kHz spec.)}.$$

The results are

b_0	-153 dB	5.0×10^{-16}	rad^2/Hz
b_{-1}	-132.5 dB	5.6×10^{-14}	rad^2/Hz
b_{-2}		(not visible)	
b_{-3}	-128.5 dB	1.4×10^{-13}	rad^2/Hz

We interpret the spectrum as follows.

1. The white phase noise is $b_0 = 5 \times 10^{-16}$ rad^2/Hz. Thus

$$P_0 = \frac{FkT}{b_0} \simeq 10 \ \mu W \quad (-20 \ \text{dBm}),$$

   assuming that $F = 1$ dB. As expected in this type of oscillator, the power is kept low for best stability.
2. The $1/f$ noise shows up in almost two decades. The Leeson effect (the slope changes by 2, $f^0 \to 1/f^2$ or $1/f \to 1/f^3$) occurs at a lower frequency. In the $1/f$ region we observe the noise of the sustaining amplifier and buffer, which adds up (subsection 3.3.2).
3. Assuming that there are four cascaded amplifiers, i.e. a sustaining amplifier and three buffers, the contribution of the sustaining amplifier is one-quarter of the total. Taking away 6 dB, we find

$$(b_{-1})_{\text{ampli}} = 1.4 \times 10^{-14} \quad (-138.5 \ \text{dB rad}^2/\text{Hz});$$

   thus $f_c = 3.2$ Hz.
4. The $1/f^3$ line crosses the flicker at $f = f_L'' = 3.2$ Hz, which is suspected to be the Leeson frequency.
5. Substituting $f_L'' = 3.2$ Hz into $Q = \nu_0/(2 f_L)$, we find $Q_s = 7.9 \times 10^5$.
6. The value $Q_s = 7.9 \times 10^5$ is far too low for it to be the resonator quality factor. The technology suggests $Q_t = 2 \times 10^6$.
7. Using $Q_t = 2 \times 10^6$ as the quality factor in $f_L = \nu_0/(2Q)$, we find $f_L = 1.25$ Hz. This indicates that the oscillator frequency flicker (b_{-3}/f^3 in the phase-noise spectrum) is not due to the Leeson effect. Rather, it is due to the $1/f$ fluctuation of the resonator's natural frequency.
8. The frequency flicker expressed as the Allan variance (Table 1.4), is

$$\sigma_y^2(\tau) = 2 \ln 2 \, h_{-1} = 2 \ln 2 \, \frac{b_{-3}}{\nu_0^2} = 1.39 \times \frac{1.4 \times 10^{-13}}{(5 \times 10^6)^2};$$

   thus

$$\sigma_y^2(\tau) \simeq 7.83 \times 10^{-27} \quad \text{and so} \quad \sigma_y(\tau) \simeq 8.85 \times 10^{-14}.$$

This is close to the specification for the lowest-noise option, given as $\sigma_y(\tau) = 1 \times 10^{-13}$ for $1 \ \text{s} \leq \tau \leq 30 \ \text{s}$.

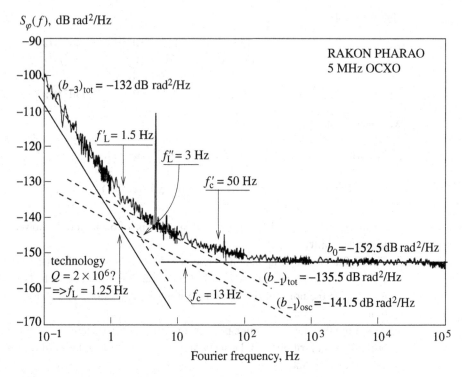

Figure 6.6 Phase-noise spectrum of the RAKON PHARAO oven-controlled quartz oscillator. Courtesy of RAKON (formerly CMAC), 2007. The interpretation, comments, and any mistakes are those of the present author.

RAKON PHARAO 5 MHz quartz oscillator

This oven-controlled quartz oscillator is manufactured in a small series of manually trimmed samples, intended as the flywheel for the PHARAO atomic fountain. In this application, a lower frequency flicker would be required, if available. Figure 6.6 shows the phase-noise spectrum of one oscillator. This spectrum is obtained as half the raw spectrum measured with two equal oscillators. In the process of selecting manually the best samples out of a small series, it is quite reasonable to trust the hypothesis that the two sections under test are almost equal.

The spectrum is again of type 2 ($f_L < f_c$), typical of low-frequency oscillators with high-Q resonators. It is fitted by the polynomial $\sum_{i=-3}^{0} b_i f^i$, with

b_0	-152.5 dB	5.62×10^{-16}	rad^2/Hz
b_{-1}	-135.5 dB	2.82×10^{-14}	rad^2/Hz
b_{-2}		(not visible)	
b_{-3}	-132 dB	6.31×10^{-14}	rad^2/Hz

and interpreted as follows.

1. The white phase noise is $b_0 = 5.62 \times 10^{-16}$ rad^2/Hz (-152.5 dB rad^2/Hz). Thus

$$P_0 = \frac{FkT}{b_0} \simeq 9.0 \; \mu\text{W} \quad (-20.5 \; \text{dBm}) \,.$$

   assuming that $F = 1$ dB. Once again, we notice that the power is kept low to achieve the best stability.

2. The phase flickering is clearly visible in almost two decades, at some 1–100 Hz. At the corner frequencies $f'_L = 1.5$ Hz and $f'_c = 50$ Hz, the difference between the experimental data and the asymptotic approximation is 3 dB, as expected.

3. The residual noise of the instrument is not shown on the plot. Yet experience suggests that the value $(b_{-1})_{\text{osc}} = 5.6 \times 10^{-14}$ rad^2/Hz (-135.5 dB) is significantly larger than the noise of a HF–VHF mixer. These elements indicate that the value $(b_{-1})_{\text{osc}} = 2.82 \times 10^{-14}$ rad^2/Hz can be trusted.

4. From the plot, the frequency flicker is $b_{-3} = 6.31 \times 10^{-14}$. We use it for a first estimate of the Leeson frequency, at which $b_{-3} f^{-3} = b_{-1} f^{-1}$: the result is $f'_L = 1.5$ Hz.

5. Let us assume that the sustaining amplifier contributes one-quarter of the total phase flickering. This is equivalent to guessing that the buffer consists of three stages (subsection 3.3.2), on the basis that the technology is similar to that of the sustaining amplifier. It follows that

$$(b_{-1})_{\text{ampli}} = 7.1 \times 10^{-15} \; \text{rad}^2/\text{Hz} \quad (-141.5 \; \text{dB rad}^2/\text{Hz}) \,;$$

   thus $f_c = 13$ Hz.

6. A provisional estimate of the Leeson frequency $f''_L \simeq 3$ Hz can now be made.

7. Using $f''_L = 3$ Hz in $f_L = v_0/(2Q)$, we get $Q_s \simeq 8.4 \times 10^5$.

8. The value $Q_s \simeq 8.4 \times 10^5$ is too low a value for a top-technology 5 MHz quartz resonator. The quality factor can be 2×10^6 or more in actual load conditions, that is, including the dissipative effect of the oscillator circuit.

9. Trusting the technology, we use $Q_t = 2 \times 10^6$. Thus

$$f_L = \frac{v_0}{2Q} \approx 1.25 \; \text{Hz} \,.$$

   This indicates that the oscillator frequency flicker $b_{-3} \simeq 6.31 \times 10^{-14}$ is due to frequency flickering in the resonator, and that the corner frequency $f''_L \simeq 3$ Hz of Fig. 6.6 is not the Leeson frequency. Instead, the Leeson effect is hidden by the resonator's instability and by the phase noise of the output buffer.

10. The flicker noise of the sustaining amplifier combined with the guessed Leeson frequency of 1.25 Hz gives a stability limit for the electronics of the oscillator. This is the solid line f^{-3} in Fig. 6.6, some 7–8 dB below the phase noise.

11. The frequency flicker expressed as the Allan variance (Table 1.4), is

$$\sigma_y^2(\tau) = 2 \ln 2 \, h_{-1} = 2 \ln 2 \, \frac{b_{-3}}{v_0^2} = 1.39 \times \frac{6.31 \times 10^{-14}}{(5 \times 10^6)^2} \,;$$

thus

$$\sigma_y^2(\tau) \simeq 3.5 \times 10^{-27} \qquad \text{and so} \qquad \sigma_y(\tau) \simeq 5.9 \times 10^{-14} .$$

FEMTO-ST LD-cut quartz oscillator (10 MHz)

Owing to the nonlinearity inherent in a quartz crystal, the resonator's natural frequency depends on the driving power. This phenomenon, referred to as the amplitude-to-frequency effect or sometimes as the isochronism defect, is of order $1.2 \times 10^{-9}/\mu W$ calculated from $\Delta \nu/(\nu_0 \Delta P)$ for the popular SC-cut crystals. Attempts to eliminate the isochronism defect led to the LD cut [42, 35], in which the residual power dependence is a few parts in $10^{-11}/\mu W$.

The oscillator analyzed in this subsection (Fig. 6.7) is a laboratory prototype; thus we know two important parameters, the quality factor and the power. This oscillator was intended to demonstrate the frequency stability of an LD-cut resonator at low Fourier frequencies, in the $1/f^3$ region. This is unfortunate for our analysis because some pieces of information were not identified as important and were discarded from the record. In particular, the white-noise level of Fig. 6.7 is inconsistent with the amplifier input power and with general experience. Keeping this in mind, we proceed with the analysis.

Figure 6.7 shows the phase-noise spectrum, measured as half the total noise obtained by comparing two equal oscillators. The plot is fitted by the polynomial $\sum_{i=-3}^{0} b_i f^i$ with

b_0	-147 dB	2×10^{-15}	rad^2/Hz
b_{-1}	-130 dB	1×10^{-13}	rad^2/Hz
b_{-2}		(not visible)	
b_{-3}	-116.6 dB	2.2×10^{-12}	rad^2/Hz

We interpret the spectrum as follows.

1. As we know the power at the amplifier input ($P_0 = 340$ μW), we can calculate the noise figure $F = b_0/(kT)$. The result is $F \simeq 167$ (22 dB), which is far too high for a HF–VHF amplifier. Consequently, the measured floor cannot be the noise floor b_0 of the oscillator. Instead, it could be due to insufficient power at the input of the noise measurement system or to other experimental problems.
2. This oscillator is a laboratory prototype intended to demonstrate the benefit of the LD cut at low Fourier frequencies. From this standpoint, it is experimentally correct to ignore a problem showing up in the white-noise region while still trusting the other measured noise coefficients.
3. The oscillator b_0 is not accessible, thus the amplifier corner frequency f_c cannot be estimated.
4. The phase flickering is clearly visible over a span of one decade. At the corner frequencies $f_L' = 5$ Hz and $f_c' = 50$ Hz, the difference between the experimental data and the asymptotic approximation is 3 dB as expected. Experience suggests that the value $(b_{-1})_\text{osc} = 10^{-13}$ can be trusted because it is sufficiently large compared with the noise of the HF–VHF mixer used to measure the oscillator phase noise.

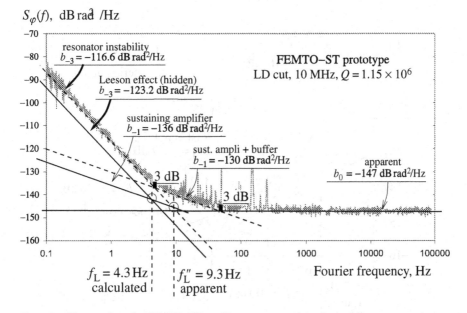

$S_\varphi(f)$, dB rad² /Hz

Figure 6.7 Phase noise of a FEMTO-ST oscillator prototype based on a LD-cut resonator; $P_0 = 340\ \mu W$, the dissipated power $= 200\ \mu W$. The interpretation, comments, and any mistakes are those of the present author.

5. We use the frequency flicker $b_{-3} = 2.2 \times 10^{-12}$ to estimate the Leeson frequency, at which $b_{-3} f^{-3} = b_{-1} f^{-1}$. A value $f_L' = 8.9$ Hz results.
6. The details of the buffer were unfortunately lost. Hence, our analysis has to rely on the hypothesis that the sustaining amplifier contributes one-quarter of the total phase flickering, as in commercial "black boxes." It follows that

$$(b_{-1})_{\text{ampli}} = 2.5 \times 10^{-14}\ .$$

7. The quality factor of the resonator in actual load conditions is known, $Q = 1.15 \times 10^6$. The calculated Leeson frequency is

$$f_L = \frac{\nu_0}{2Q} = 4.35\ \text{Hz}\ .$$

This value suffers from the uncertainty involved in guessing the phase flickering of the sustaining amplifier as a fraction of the total phase flickering.
8. The frequency at which $b_{-3}/f^3 = b_{-1}/f$ is $f_L'' = 9.3$ Hz. This value is significantly higher than the calculated Leeson frequency $f_L = 4.35$ Hz. This fact provides evidence that the Leeson effect is not visible on the spectrum and that the $1/f^3$ line is the frequency flickering of the resonator.
9. The frequency flicker expressed as the Allan variance (Table 1.4) is

$$\sigma_y^2(\tau) = 2\ln 2\ h_{-1} = 2\ln 2\ \frac{b_{-3}}{\nu_0^2} = 2 \times \frac{2.2 \times 10^{-12}}{(10 \times 10^6)^2}\ ;$$

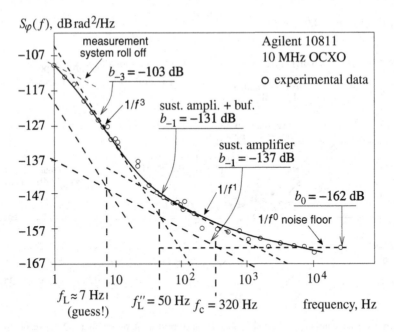

Figure 6.8 Phase noise of the Agilent 10811 10 MHz oscillator [15]. Used with the permission of Agilent Technologies, 2007. The interpretation, comments, and any mistakes are those of the present author.

hence

$$\sigma_y^2(\tau) \simeq 3 \times 10^{-26} \quad \text{and so} \quad \sigma_y(\tau) \simeq 1.74 \times 10^{-13} .$$

Agilent 10811 quartz (10 MHz)

This oven-controlled quartz oscillator has been the high-stability workhorse for a number of Agilent – and formerly Hewlett Packard – electronic instruments. It is no longer supported by Agilent but the manual is available on a website maintained by the community of users [51].

Figure 6.8 shows the phase-noise spectrum. The spectrum is fitted by the polynomial $\sum_{i=-3}^{0} b_i f^i$, with

b_0	-162 dB	6.3×10^{-17}	rad^2/Hz
b_{-1}	-131 dB	7.9×10^{-13}	rad^2/Hz
b_{-2}		(not visible)	
b_{-3}	-103 dB	5.0×10^{-11}	rad^2/Hz

We interpret the spectrum as follows.

1. Under the usual assumption that $F = 1$ dB, the white phase noise yields

$$P_0 = \frac{FkT}{b_0} \simeq 80 \ \mu W \quad (-11 \ \text{dBm}) .$$

The moderate operating power helps in achieving high long-term stability.

2. This oscillator is intended for regular production rather than for special applications. It is therefore reasonable that the phase noise is somewhat higher than that of some of the special units already discussed. This fact makes its measurement easier, which further encourages us to trust the spectrum of Fig. 6.8.
3. At the corner frequencies, 25 Hz and 1.3 kHz, the difference between the experimental data and the asymptotic approximation is 3 dB, as expected.
4. In the 50–300 Hz region, the phase noise is of the $1/f$ type. This includes sustaining amplifier and buffer. Taking away 6 dB (three buffer stages), the estimated flickering of the sustaining amplifier is

$$(b_{-1})_{\text{ampli}} = 2 \times 10^{-14} \text{ rad}^2/\text{Hz} \qquad (-137 \text{ dB rad}^2/\text{Hz}) \, ;$$

thus $f_c = 320$ Hz.
5. The oscillator $1/f^3$ noise crosses the sustaining-amplifier noise $(b_{-1})_{\text{ampli}} f^{-1}$ at $f''_L \simeq 50$ Hz. Whether f''_L is the Leeson frequency is still to be decided.
6. Using $f''_L = 50$ Hz in $f_L = v_0/(2Q)$, we get $Q_s \simeq 1 \times 10^5$.
7. The value $Q_s \simeq 1 \times 10^5$ is far too low for it to be the resonator quality factor. To be conservative, we believe that the quality factor Q_t cannot be lower than 7×10^5. Thus

$$f_L = \frac{v_0}{2Q} \approx 7 \text{ Hz} \, .$$

Accordingly, the oscillator frequency flicker $b_{-3} \simeq 5 \times 10^{-11}$ is due to the frequency flickering in the resonator, while the Leeson effect is some 17 dB lower.
8. The frequency flicker expressed as the Allan variance (Table 1.4) is

$$\sigma_y^2(\tau) = 2 \ln 2 \, h_{-1} = 2 \ln 2 \, \frac{b_{-3}}{v_0^2} = 1.39 \times \frac{5 \times 10^{-11}}{(10 \times 10^6)^2} \, ;$$

thus

$$\sigma_y^2(\tau) \simeq 6.9 \times 10^{-25} \qquad \text{and so} \qquad \sigma_y(\tau) \simeq 8.3 \times 10^{-13} \, .$$

The $1/f$ noise of the sustaining amplifier deserves a final comment. This noise is inferred by subtracting 6 dB from the $1/f$ region of the output spectrum, in the same way as for other oscillators analyzed in this chapter. It is surprising, however, to notice that the numerical value of this $1/f$ noise is similar to that of the other oscillators, most of which are intended for far more demanding – and far more expensive – applications.

Agilent noise-degeneration oscillator (10 MHz)

This oscillator, tested at the Agilent Laboratories, [60], includes a noise-degeneration system based on the ideas of subsection 2.5.3, which reduces the flicker noise of the amplifier. Figure 6.9 shows the phase-noise spectrum. The plot is fitted by the

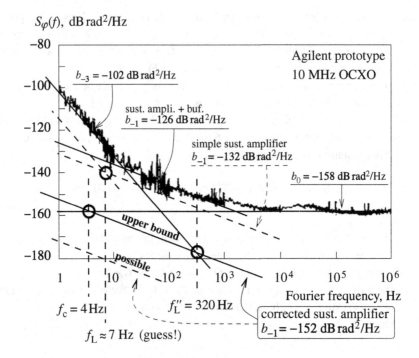

$S_\varphi(f)$, dB rad^2/Hz

Figure 6.9 Phase noise of the balanced-bridge quartz oscillator tested at Agilent Laboratories [60]. Used with the permission of the IEEE, 2007. The interpretation, comments, and any mistakes are those of the present author.

polynomial $\sum_{i=-3}^{0} b_i f^i$, with

b_0	-158 dB	1.6×10^{-16}	rad^2/Hz
b_{-1}	-126 dB	2.5×10^{-13}	rad^2/Hz
b_{-2}		(not visible)	
b_{-3}	-102 dB	6.3×10^{-11}	rad^2/Hz

We interpret the spectrum as follows.

1. The noise-degeneration amplifier exhibits low flicker, at the expense of a higher noise figure than that of a simple amplifier. This higher noise figure is due to the need to split the input signal into two branches, for power amplification and for noise correction of the power amplifier (subsection 2.5.3). We assume that $F = 5$ dB; this is inferred by accounting for the 3 dB loss inherent in power splitting, plus the 1 dB dissipative loss and 1 dB noise figure of the internal amplifiers.
2. The white phase noise is $b_0 = 1.6 \times 10^{-16}$ rad^2/Hz (-158 dB rad^2/Hz). Thus

$$P_0 = \frac{FkT}{b_0} \simeq 80 \ \mu\text{W} \qquad (-11 \text{ dBm}).$$

This power level is consistent with one's general experience of 5–10 MHz oscillators, in which the power is kept low for best stability.

3. The visible corner frequency, at which the f^{-3} noise crosses the f^{-1} noise, is

$$f_L' = \sqrt{\frac{b_{-3}}{(b_{-1})_{\text{tot}}}} \simeq 16 \text{ Hz} .$$

4. Subsection 3.3.2 indicates that the effect of the output buffer cannot be neglected if phase flickering shows up in the phase-noise spectrum. Elsewhere we have assumed that there are three buffer stages between the sustaining amplifier and the output, the $1/f$ noise of each stage being equal to that of the sustaining amplifier and the flicker noise of the sustaining amplifier being $(b_{-1})_{\text{ampli}} = \frac{1}{4}(b_{-1})_{\text{tot}}$. Thus

$$(b_{-1})_{\text{ampli}} = 6.3 \times 10^{-14} \text{ rad}^2/\text{Hz} \qquad (-132 \text{ dB rad}^2/\text{Hz})$$

$$\text{(for a simple amplifier, not for this oscillator)} .$$

   This is indicated as a broken line in Fig. 6.9.
5. Owing to feedback, the noise-degeneration amplifier exhibits low flicker, limited by the internal phase measurement system. Let us assume that the $1/f$ noise improves on a non-corrected amplifier by 20 dB, as a conservative value. Hence, an upper bound of the amplifier noise is

$$(b_{-1})_{\text{ampli}} = 6.3 \times 10^{-16} \text{ rad}^2/\text{Hz} \qquad (-152 \text{ dB rad}^2/\text{Hz})$$

$$\text{(noise corrected amplifier)} .$$

   In Fig. 6.9, this is the solid line 20 dB lower than the noise of a simple amplifier and 26 dB lower than the oscillator output $1/f$ noise.
6. The corner frequency of the sustaining amplifier is

$$f_c = \frac{b_0}{(b_{-1})_{\text{ampli}}} \simeq 4 \text{ Hz} .$$

7. The visible f^{-3} noise crosses the estimated f^{-1} noise of the sustaining amplifier at

$$f_L'' = \sqrt{\frac{b_{-3}}{(b_{-1})_{\text{ampli}}}} \simeq 320 \text{ Hz} .$$

8. Taking $f_L'' = 320$ Hz as the Leeson frequency and using $f_L = v_0/(2Q)$ we get $Q_s \simeq 1.6 \times 10^4$.
9. The value $Q_s \simeq 1.6 \times 10^4$ is far too low a value for a state-of-the-art oscillator. The literature suggests that the product $v_0 Q$ is between 10^{13} and 2×10^{13} (unloaded), reduced by a factor 0.5–0.8 when loaded.
10. The actual Q value is higher than 1.6×10^4. This means that the Leeson effect is hidden, and that we are not able to calculate Q. Hence we can only guess the value of Q, relying on experience and on the literature.
11. If we admit $Q = 10^6$, reduced to $Q = 7 \times 10^5$ (loaded) by the dissipative effect of the amplifier input, we get

$$f_L = \frac{v_0}{2Q} \approx 7.1 \text{ Hz} .$$

This would indicate that the oscillator frequency flicker $b_{-3} \simeq 6.3 \times 10^{-11}$ is due to frequency flickering in the resonator, not the Leeson effect. Accordingly, the corner visible in Fig. 6.9 at 16 Hz will not occur at the Leeson frequency.

12. Using

$$(b_{-3})_{\text{osc}} = \frac{(b_{-1})_{\text{ampli}}}{f_L^2} ,$$

we find that the contribution of the Leeson effect to the oscillator $1/f^3$ noise is

$$(b_{-3})_{\text{osc}} \simeq 3.2 \times 10^{-16} \text{ rad}^2/\text{Hz} \qquad (-135 \text{ dB rad}^2/\text{Hz}) .$$

This value is about 33 dB lower than the actual b_{-3} coefficient of the oscillator noise.

13. With reference to Fig. 3.12, we note that the spectrum is of type 1 B, for which $f_c < f_L$. It is worth mentioning that we have chosen an upper bound for the amplifier $1/f$ noise. A value lower than -152 dB rad^2/Hz results in a value of f_c lower than 4 Hz, which reinforces the conclusion that $f_c < f_L$.

14. The frequency flicker expressed as the Allan variance (Table 1.4), is

$$\sigma_y^2(\tau) = 2 \ln 2 \, h_{-1} = 2 \ln 2 \, \frac{b_{-3}}{v_0^2} = 1.39 \times \frac{6.3 \times 10^{-11}}{(10^7)^2} ;$$

thus

$$\sigma_y^2(\tau) \simeq 8.8 \times 10^{-25} \qquad \text{and so} \qquad \sigma_y(\tau) \simeq 9.3 \times 10^{-13} .$$

There is some disagreement between this value and the value $\sigma_y = 2 \times 10^{-12}$ at $\tau = 1$ s given by [60, Table I]. This disagreement is present in the article cited and is independent of our interpretation.

15. The Allan variance calculated only on the basis of the Leeson effect, thus with a noise-free resonator but accounting for an upper bound of the amplifier $1/f$ noise, is

$$\sigma_y^2(\tau) = 2 \ln 2 \, h_{-1} = 2 \ln 2 \, \frac{b_{-3}}{v_0^2} = 1.39 \times \frac{3.2 \times 10^{-14}}{(10^7)^2} ;$$

thus

$$\sigma_y(\tau) \simeq 2.1 \times 10^{-14} \qquad \text{(noise-free resonator)} .$$

Of course, this high stability is out of reach for a quartz oscillator.

When this project was presented at the 1999 EFTF/FCS joint meeting in Besançon and published in [60], a number of researchers and engineers then expected a new low-noise oscillator from Agilent. This did not happen. The reasons are of course confidential. Nonetheless, we can infer that the new and smart sustaining amplifier was an unnecessary improvement on a well-designed traditional amplifier, since the oscillator stability is in any case limited by the fluctuations of the resonator's natural frequency.

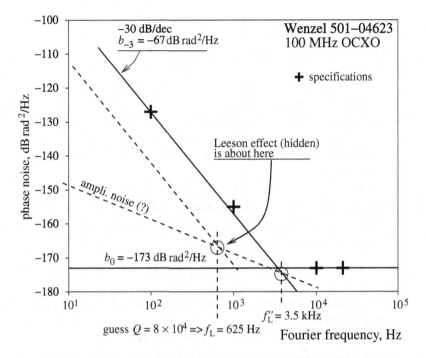

Figure 6.10 Phase noise of the 100 MHz oscillator Wenzel 501-04623. The numerical values are taken from the manufacturer's website. The plot, interpretation, comments, and any mistakes are those of the present author.

Wenzel 501-04623 (100 MHz SC-cut quartz)

The Wenzel 501-04623 oscillator is chosen as an example because of its outstandingly low phase-noise floor and because of its high carrier frequency (100 MHz). These two features are necessary for low noise after frequency multiplication. A table of phase-noise values is given by the manufacturer, rather than the complete spectrum. These values, used in plotting the spectrum (Fig. 6.10), are:

b_0	-173 dB	5×10^{-18}	$\mathrm{rad}^2/\mathrm{Hz}$
b_{-1}		(not visible)	
b_{-2}		(not visible)	
b_{-3}	-67 dB	2×10^{-7}	$\mathrm{rad}^2/\mathrm{Hz}$

An interpretation is as follows.

1. The white phase noise is $b_0 = 5 \times 10^{-17}$ $\mathrm{rad}^2/\mathrm{Hz}$ (-173 dB $\mathrm{rad}^2/\mathrm{Hz}$). Thus

$$P_0 = \frac{FkT}{b_0} \simeq 1\ \mathrm{mW} \qquad (0\ \mathrm{dBm}),$$

assuming that $F = 1$ dB. This relatively large power is necessary for low short-term noise, although it may be detrimental to the medium-term and long-term stability.

2. The terms $b_{-2}f^{-2}$ and $b_{-1}f^{-1}$ must be negligible because there is no room for them in the region between the two specified points, 1 kHz and 10 kHz. Confirmation

comes from the measurement of some samples. Consequently, the point at which the slope changes by 2 (either f^0 to f^{-2} or f^{-1} to f^{-3}) is not visible. This indicates that the Leeson effect is hidden by the frequency flicker of the resonator.

3. Of course, the corner point where the frequency flicker crosses the white noise ($b_{-3}f^{-3} = b_0$), which occurs at $f = 3$ kHz, is not the Leeson frequency.

4. A further consequence of the absence of a visible $1/f$ region in the spectrum is that we have no data from which to infer the $1/f$ noise of the sustaining amplifier. The flicker noise of a state-of-the-art HF–VHF amplifier is between -135 and -140 dB rad^2/Hz. Thus, we will take $(b_{-1})_{\text{ampli}} \approx 1.6 \times 10^{-14}$ rad^2/Hz (-138 dB rad^2/Hz) as a plausible value.

5. As the Leeson effect is hidden, we are not able to calculate Q. Relying upon the literature, we guess that a high-stability 100 MHz quartz resonator will have a quality factor of 1.2×10^5, reduced to $Q = 8 \times 10^4$ in actual load conditions. Accordingly, the Leeson frequency is

$$f_L = \frac{\nu_0}{2Q} \simeq 625 \text{ Hz} .$$

6. The sustaining-amplifier $1/f$ noise, combined with the Leeson frequency, gives the stability limit of the electronics of the oscillator. This is the broken line $6.2 \times 10^{-9}/f^3$ in Fig. 6.10 that is 15.1 dB below the phase noise.

7. The frequency flicker expressed as the Allan variance (Table 1.4), is

$$\sigma_y^2(\tau) = 2 \ln 2 \, h_{-1} = 2 \ln 2 \frac{b_{-3}}{\nu_0^2} = 1.39 \times \frac{2 \times 10^{-7}}{(100 \times 10^6)^2} ;$$

thus

$$\sigma_y^2(\tau) \simeq 2.8 \times 10^{-23} \qquad \text{and so} \qquad \sigma_y(\tau) \simeq 5.3 \times 10^{-12} .$$

6.5   The origin of instability in quartz oscillators

Let us introduce the noise ratio R, defined as

$$R = \sqrt{\frac{(b_{-3})_{\text{tot}}}{(b_{-3})_L}} . \tag{6.7}$$

This is the total frequency flicker at the amplifier output divided by the frequency flicker due to the sustaining amplifier via the Leeson effect. Although not mentioned explicitly, R can be easily identified in all the phase-noise spectra of this chapter. As a consequence of (6.7), the parameter R is also given by

$$R \approx \frac{f_L''}{f_L} , \tag{6.8}$$

where f_L'' is the frequency at which the continuation of the oscillator $1/f^3$ noise crosses the $1/f$ noise of the sustaining amplifier (Fig. 6.2). The approximation is due to the uncertainty in identifying or guessing f_L''. Additionally R is related to the quality factor by

$$R \approx \frac{Q}{Q_s} \approx \frac{Q_t}{Q_s},$$ (6.9)

where Q_s is calculated using f_L'' as if it were the Leeson frequency and where the true Q is replaced by Q_t, that is, the most plausible value based on the available technology. This can be seen by introducing the power law (1.70) into the Leeson formula, written in terms of ν_0 and Q, (3.19), or in terms of f_L, (3.21). The above-mentioned approximation derives from f_L'', which enters in Q_s, and then from Q_t.

Finally, R can be expressed as the stability ratio

$$R = \frac{(\sigma_y)_{osc}}{(\sigma_y)_L} \quad \text{(flicker floor)},$$ (6.10)

that is, the Allan-deviation floor of the complete oscillator calculated from the observed $1/f^3$ phase noise, divided by the floor calculated on the basis of the Leeson effect only, assuming that the resonator is free from noise. Equation (6.10) is derived from (6.7) using Table 1.4 or Fig. 1.8. The parameter R states the *poorness* of the actual oscillator, as compared with the same oscillator if it were governed only by the Leeson effect and had no resonator fluctuations. This statement can be rephrased by saying that R is the *goodness* of the sustaining amplifier compared with the oscillator fluctuations. Thus:

$R = 1$ (0 dB) indicates that the oscillator f^{-3} phase noise comes from the Leeson effect and that the resonator fluctuations are negligible;

$R = \sqrt{2}$ (3 dB) indicates that the resonator $1/f$ fluctuations and the Leeson effect contribute equally to the oscillator noise;

$R \gg \sqrt{2}$ (3 dB) indicates that resonator instability is the main cause of the f^{-3} phase noise.

Table 6.1 gives a comparison of the results. Some 5–10 MHz oscillators are similar, to the extent that all the available technology is employed in order to get the highest stability. In all cases we have analyzed, we find R values of the order of 10 dB, with a minimum of 6.6 dB. This means that the Leeson effect is hidden below the frequency fluctuation of the resonator. The Agilent 10811 is closer to routine in production, and probably closer to a cost–performance trade-off, than the others; thus understanding this oscillator is more difficult. Nonetheless, in this case the value of Q_s is so low that there is no doubt that it cannot be the resonator's quality factor. The Wenzel 501-04623 (100 MHz) is clearly designed for minimal phase noise in the white region. This requires a high driving power, so it is not surprising that the resonator stability is impaired.

The examples shown above indicate that an oscillator's frequency flickering is chiefly due to the fluctuations in the resonator's natural frequency. This fact has a number of implications, some of which are unexpected.

Table 6.1 Estimated parameters of some ultra-stable oscillators. Adapted from [83] and used with the permission of the IEEE, 2007

Oscillator	ν_0	b_{-3}, tot.	b_{-1}, tot.	b_{-1}, ampli.	f'_L	f''_L	Q_s	Q_t	f_L	$(b_{-3})_L$	R	Ref(s).
Oscilloquartz 8600[a]	5	−124.0	−131.0	−137.0	2.24	4.5	5.6×10^5	1.8×10^6	1.4	−134.1	10.1	[99, 73]
Oscilloquartz 8607[a]	5	−128.5	−132.5	−138.5	1.6	3.2	7.9×10^5	2×10^6	1.25	−136.5	8.1	[99, 73]
CMAC Pharao[b]	5	−132.0	−135.5	−141.1	1.5	3	8.4×10^5	2×10^6	1.25	−139.6	7.6	[17, 16, 21]
FEMTO-ST LD protot.[c]	10	−116.6	−130.0	−136.0	4.7	9.3	5.4×10^5	1.15×10^6	4.3	−123.2	6.6	[35]
Agilent 10811[d]	10	−103.0	−131.0	−137.0	25	50	1×10^5	7×10^5	7.1	−119.9	16.9	[15]
Agilent prototype[e]	10	−102.0	−126.0	−152.0	16	320	1.6×10^4	7×10^5	7.1	−134.9	32.9	[60]
Wenzel 501-04623[f]	100	−67.0	−132 ?	−138 ?	1800	3500	1.4×10^4	8×10^4	625	−79.1	15.1	[107]
unit	MHz	dB rad²/Hz			Hz	Hz	(none)	(none)	Hz	dB rad²/Hz	dB	

[a] The data are from specifications with full options about low noise and high stability.
[b] Measured by RAKON on a unit; RAKON confirmed $2 \times 10^6 < Q < 2.8 \times 10^6$ in actual conditions.
[c] LD-cut, built, and measured in our laboratory, yet by a different team. All design parameters are known and hence Q_t.
[d] Measured by Hewlett Packard (now Agilent) on a sample.
[e] Implements a bridge scheme for the degeneration of the amplifier noise. Same resonator as that of the Agilent 10811.
[f] The data are from specifications.

1. An oscillator's frequency flickering can only be improved by improving the resonator.
2. As a consequence, there is no way of improving the frequency flickering through innovative design of the sustaining amplifier. To this extent, the electronics is at a dead end.
3. The analysis of the 10811 oscillator confirms the above conclusion. As a matter of fact, the noise of the sustaining amplifier is similar to that of higher-stability oscillators. We do not know anything about the cost-performance trade-off in the design of this oscillator. Nonetheless, we notice that the $1/f$ noise of the sustaining amplifier is not increased by a lower budget.
4. Surprisingly, in ultra-stable oscillators the $1/f$ phase noise of the output buffer shows up, and perhaps also the $1/f$ noise of the sustaining amplifier, which can be annoying in the 0.1–1 s region of the Allan deviation. This fact leaves room for innovative electronic design.
5. After observing that simple and relatively cheap radio-electronics is sufficient for the stability of a high-end oscillator to be limited by the resonator, it may be pointed out that the stability of lower-rank oscillators is clearly dominated by the resonator.
6. Finally, we recall that this analysis is based only on the phase noise affecting the amplifier's forward gain. The noise originating in the impedance fluctuation of the amplifier transferred to the resonator's natural frequency (subsection 3.3.4) has not been taken into account. This noise phenomenon, whose existence is even not mentioned in the literature, deserves theoretical development and experiments.

6.6 Microwave oscillators

Microwave oscillators differ radically from quartz oscillators in that the $1/f$ frequency noise is governed by the amplifier's $1/f$ phase noise through the Leeson effect. This is related to three differences between microwave oscillators and quartz oscillators.

1. The volume of energy confinement is significantly larger in microwave resonators than in quartz resonators.
2. Excluding cryogenic units, the quality factor of a microwave resonator is usually lower by a factor 10–10^3 than that of a quartz resonator.
3. Microwave amplifiers show larger $1/f$ noise than RF amplifiers.

Additionally, circulators and isolators are available only at microwave frequencies, while the reverse isolation of a microwave amplifier is usually rather poor.

This section gives a number of worked-out examples of microwave oscillators.

Miteq DRO mod. D-210B

Figure 6.11 shows the phase-noise spectrum of the dielectric resonator oscillator (DRO) Miteq D-210B, taken from the device data sheet. The plot is fitted by the

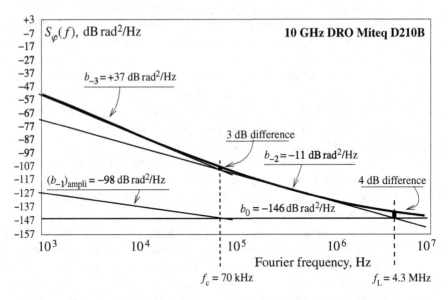

Figure 6.11 Phase noise of the 10 GHz DRO Miteq D210B. Used with the permission of Miteq Inc., 2007. The interpretation, comments, and any mistakes are those of the present author.

polynomial $\sum_{i=-3}^{0} b_i f^i$, with

b_0	-146 dB	2.5×10^{-15}	rad^2/Hz
b_{-1}		(not visible)	
b_{-2}	-11 dB	7.9×10^{-2}	rad^2/Hz
b_{-3}	$+37$ dB	5.0×10^3	rad^2/Hz

This indicates that the spectrum is of the type 1A of Fig. 3.12.

One might be tempted to fit the spectrum with a smaller b_0 (say, -147 dB rad^2/Hz) and to add a term $b_{-1}f^{-1}$ tangent to the curve at $f \approx 2$ MHz. In this case the spectrum would be of type 1B, which contains the signature of the output buffer. We will discard this interpretation, however, because then the noise of the output buffer would be $b_{-1} \approx 10^{-8}$ rad^2/Hz (-80 dB rad^2/Hz), which is too high for a microwave amplifier (Table 2.1).

The spectrum gives the following indications.

1. The coefficient b_0 derives from the amplifier noise FkT referred to the carrier power P_0 at the input of the sustaining amplifier, that is, $b_0 = FkT/P_0$. Assuming that the noise figure is $F = 1$ dB, and thus $FkT = 5.1 \times 10^{-21}$ rad^2/Hz (-173 dB rad^2/Hz), it follows that $P_0 = FkT/b_0 = 2$ μW (-27 dBm).
2. However arbitrary the assumption $F = 1$ dB may seem, it is representative of actual microwave amplifiers. Depending on bandwidth and technology, the noise figure of a "good" amplifier is between 0.5 dB and 2 dB. In this range, we find P_0 between 1.8 μW and 2.5 μW.

3. The spectrum changes slope from f^0 to f^{-2} at the Leeson frequency $f_L \simeq 4.3$ MHz. At this frequency, the asymptotic approximation is some 4 dB lower than the measured spectrum, instead of the expected 3 dB. This discrepancy is tolerable.

4. From $f_L \simeq 4.3$ MHz, it follows that $Q = \nu_0/(2f_L) \simeq 1160$, which is quite plausible for a dielectric resonator.

5. The corner point at which the slope changes from -2 to -3 is 70 kHz. This is the corner frequency f_c of the amplifier, at which it holds that $(b_{-1})_{\text{ampli}} f^{-1} = (b_0)_{\text{ampli}}$. Hence $(b_{-1})_{\text{ampli}} = 1.8 \times 10^{-10}$ rad^2/Hz (-98 dB rad^2/Hz).

6. The flicker-frequency coefficient $b_{-3} = 5 \times 10^3$ rad^2/Hz ($+37$ dB rad^2/Hz).

7. The white and flicker frequency noise, transformed into the Allan variance (Table 1.4), is

$$
\sigma_y^2(\tau) = \frac{h_0}{2\tau} + 2\ln 2\, h_{-1}
$$

$$
= \frac{b_{-2}}{\nu_0^2}\frac{1}{2\tau} + 2\ln 2\,\frac{b_{-3}}{\nu_0^2}
$$

$$
\simeq \frac{7.9 \times 10^{-2}}{2 \times (10^{10})^2}\frac{1}{\tau} + 1.39 \times \frac{5 \times 10^3}{(10^{10})^2}\ ;
$$

thus

$$
\sigma_y^2(\tau) \simeq \frac{4 \times 10^{-22}}{\tau} + 7.0 \times 10^{-17}
$$

and[1]

$$
\sigma_y(\tau) \asymp \frac{2 \times 10^{-11}}{\sqrt{\tau}} + 8.3 \times 10^{-9}\ .
$$

Finally, one should note that the oscillator flicker shows up in the 1–100 kHz region. Common sense suggests that temperature and other environmental fluctuations have no effect on this time scale and that the flickering of the dielectric constant in the resonator will not exceed the amplifier noise. Consequently, in this region the oscillator flicker is due to the amplifier, through the Leeson effect, rather than to the resonator.

Poseidon DRO-10.4-FR (10.4 GHz)

The Poseidon DRO-10.4-FR is another example of an oscillator based on a dielectric resonator. Figure 6.12 shows the phase-noise spectrum, from a preliminary data sheet. The unusually low noise raises the question whether this DRO is a technology demonstrator or a regular product. The power spectral density $S_\varphi(f)$ is fitted by the polynomial $\sum_{i=-3}^{0} b_i f^i$, with

---

[1] The symbol $\asymp$ means *asymptotically equal*.

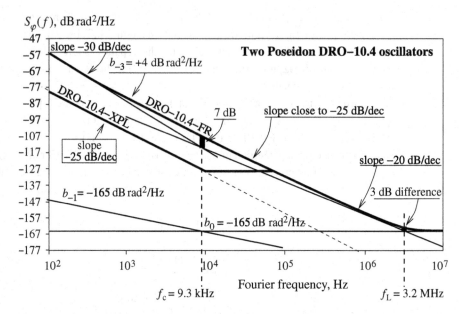

$S_\varphi(f)$, dB rad²/Hz

Figure 6.12 Poseidon DRO 10.4-FR. Used with the permission of Poseidon Scientific Instruments, 2007. The interpretation, comments and any mistakes are those of the present author.

b_0	-165 dB	3.2×10^{-17}	rad²/Hz
b_{-1}		(not visible)	
b_{-2}	-35 dB	3.2×10^{-4}	rad²/Hz
b_{-3}	$+4$ dB	2.5	rad²/Hz

Once again the spectrum is of the type 1A (Fig. 3.12) typical of microwave oscillators. Yet the discrepancy with respect to the theoretical model is larger than in the case of the Miteq oscillator. The spectrum gives the following indications.

1. According to the manufacturer's data sheet, the spectrum results from measurements on two DRO-10.4 oscillators. It is common practice to compare two oscillators in this way and to refer the spectrum to a single oscillator after taking away 3 dB from the raw data. This is legitimate under the quite realistic assumption that the two oscillators are equal and independent and thus that each contributes half the measured spectrum.

2. The white phase noise is $b_0 = 3.2 \times 10^{-17}$ rad²/Hz (-165 dB rad²/Hz). Thus $P_0 = FkT/b_0 \simeq 160$ μW (-8 dBm), assuming that $F = 1$ dB.

3. The Leeson frequency is $f_L = 3.2$ MHz. Accordingly, the quality factor $Q = \nu_0/(2f_L) \simeq 1625$, which is reasonable for a dielectric resonator.

4. In a type-1A spectrum it holds that $b_{-2}f^{-2} = b_0$ at $f = f_L$. Thus, the white frequency noise is $b_{-2} \simeq 3.2 \times 10^{-4}$ rad²/Hz (-35 dB rad²/Hz).

5. There is some discrepancy between the Leeson model and the true spectrum. In the region from 2 kHz to 200 kHz the spectrum seems to be close to a line of slope $f^{-5/2}$ rather than f^{-3} or f^{-2}. At the present time this discrepancy (up to 4 dB at $f \approx 10$ kHz) is unexplained.

6. The corner frequency of the amplifier (i.e. the frequency at which the oscillator spectrum changes from f^{-2} to f^{-3}) is $f_c = 9.3$ kHz. Accordingly, the phase-noise spectrum of the amplifier, on the left-hand side of $f = f_c$, is $(b_{-1})_{\text{ampli}} = b_0 f_c = 2.9 \times 10^{-13}$ rad^2/Hz (-125 dB rad^2/Hz).

7. The amplifier flickering, 5 dB lower than the best value of Table 2.1, is surprisingly low for a microwave amplifier within a commercial product. Such a low noise could be obtained from SiGe technology, with a single-stage amplifier employing a large-volume transistor, or from a feedback or feedforward noise-degeneration scheme. A noise-degeneration scheme seems incompatible with the size of the packaged oscillator, yet nothing definite can be inferred on the basis of the available information.

8. The flicker frequency coefficient is $b_{-3} = 2.5$ rad^2/Hz ($+4$ dB rad^2/Hz).

9. The white and flicker frequency noise, transformed into the Allan variance (Table 1.4), is

$$\sigma_y^2(\tau) = \frac{h_0}{2\tau} + 2 \ln 2 \, h_{-1}$$

$$= \frac{b_{-2}}{v_0^2} \frac{1}{2\tau} + 2 \ln 2 \frac{b_{-3}}{v_0^2}$$

$$\simeq \frac{2.5}{2 \times (10.4 \times 10^9)^2} \frac{1}{\tau} + 1.39 \times \frac{5 \times 10^3}{(10.4 \times 10^9)^2} \, ;$$

thus

$$\sigma_y^2(\tau) \simeq \frac{1.5 \times 10^{-24}}{\tau} + 3.2 \times 10^{-20}$$

and so

$$\sigma_y(\tau) \asymp \frac{1.2 \times 10^{-12}}{\sqrt{\tau}} + 1.8 \times 10^{-10} \, .$$

Figure 6.12 also reports the phase-noise spectrum of the DRO-10.4-XPL oscillator, which is a different version of the same base design intended for phase-locked loops. Below a cutoff frequency of about 70 kHz, this oscillator is locked to an external reference. In this condition, the spectrum is chiefly determined by the external reference and gives little information on the oscillator. Nonetheless, it is to be noted that the spectrum is proportional to $f^{-5/2}$ (i.e., -25 dB/decade) below 10 kHz, the loop cutoff frequency. This might be the result of an experimental problem in the measurement or the signature of a fractional-order control system, like that proposed in [24].

Poseidon Shoebox (10 GHz sapphire resonator)

The Poseidon Shoebox makes use of a sapphire whispering-gallery (WG) resonator and a bridge noise-degeneration amplifier. The result is a low-noise oscillator intended for high short-term-stability applications. Valuable information about this type of oscillator or resonator is found in [67] and [59].

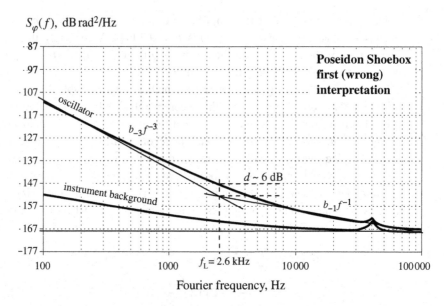

Figure 6.13 First attempt to interpret the phase-noise spectrum of the Poseidon Shoebox. Used with the permission of Poseidon Scientific Instruments, 2007. The interpretation, comments, and mistakes are those of the present author.

Figure 6.13 shows the phase-noise spectrum with a tentative interpretation. The spectrum seems to be of type 2 (Fig. 3.11), with $f_L < f_c$. Qualitatively, this is consistent with the fact that the WG resonator features a high Q value. Yet this interpretation suffers from three problems.

1. From $Q = v_0/(2 f_L)$ we get $Q \approx 1.9 \times 10^6$ at $v_0 = 10$ GHz. This value is incompatible with the dielectric loss of the sapphire. For comparison, the typical quality factor of a 10 GHz WG resonator of 5 cm diameter and 2.5–3 cm thickness is given in the following table:

Q	Temperature
2×10^5	295 K, room temperature
3×10^7	77 K, liquid N
4×10^9	4.2 K, liquid He

The resonator size cannot be reduced significantly without impairing the quality factor. The dielectric loss of the sapphire is a reproducible function of temperature. Thus, the resonator loss depends on the space distribution of the electric field, i.e., on the mode, in a narrow range.

2. Cooling to about 50 K below room temperature is possible with Peltier cells. Yet a quality factor of 1.9×10^6 cannot be obtained with this moderate cooling. A power consumption of 35 W (steady state, at 25 °C) is too low for this technology and

resonator size. Moreover, the oscillator size and weight (3 dm^3 and less than 7 kg) are incompatible with the presence of a cryo-generator. Finally, such an optical cooling technology [30, 28] is still not available and was definitely out of question when this oscillator was designed. These issues indicate that the resonator is intended to be used at room temperature.

3. The phase flicker of the sustaining amplifier is $b_{-1}f^{-1}$ with $b_{-1} \approx 10^{-12}$ rad^2/Hz. What is seen on the plot is -160 dB rad^2/Hz at $f = 10$ kHz with slope f^{-1} corresponding to -120 dB rad^2/Hz at 1 Hz. This value is too high for a sophisticated amplifier that makes use of the bridge noise-correction technique.

4. At $f = f_L$, where $b_{-3}f^{-3} = b_{-1}f^{-1}$, the true spectrum differs from the asymptotic approximation by 6 dB instead of 3 dB. This discrepancy is an additional reason to reject the interpretation of Fig. 6.13.

The above difficulties lead us to understand that the spectrum is actually of type 2A (Fig. 3.12), in which the flicker noise of the output buffer shows up in the region around f_L. There follows a completely new interpretation, shown in Fig. 6.14.

1. As usual, we start from the white noise floor. On the figure we observe that $b_0 = 1.3 \times 10^{-17}$ rad^2/Hz (-169 dB rad^2/Hz). This is ascribed to the sustaining amplifier.

2. The sustaining amplifier makes use of a bridge noise-degeneration circuit to reduce the flicker noise. In this circuit there are two amplifiers: the first amplifies the input signal and the second is used to correct the noise of the first, at the output of a

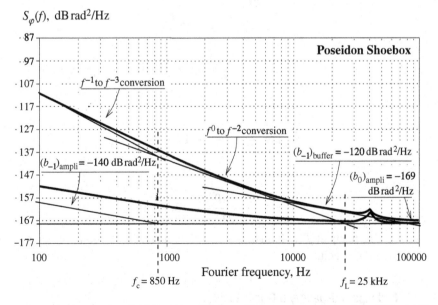

Figure 6.14 Phase noise of the Poseidon Shoebox. The same spectrum as that of Fig. 6.13 but with a revised interpretation. Used with the permission of Poseidon Scientific Instruments, 2007. The interpretation, comments, and any mistakes are those of the present author.

carrier-suppression circuit. It does this by means of a feedback circuit. We estimate a noise figure $F = 5$ dB, which results from the intrinsic loss of the power splitter at the input of the noise-corrected amplifier (3 dB), from the resistive loss of the power splitter and of the lines (1 dB), and from the noise figure of the second amplifier (1 dB).

3. From $b_0 = FkT/P_0$ we get $P_0 = 1$ mW (0 dBm).

4. The phase flickering of the output buffer shows up on the right-hand part of the spectrum, at 10^5 Hz and beyond. The noise coefficient is $(b_{-1})_{buf} \simeq 10^{-12}$ rad^2/Hz (-120 dB rad^2/Hz). The output buffer is a good microwave amplifier.

5. After removing the buffer phase noise, the white frequency noise $b_{-2}f^{-2}$ of the oscillator may be clearly identified. The decade centered at 4 kHz is used to find $b_{-2} = 7.9 \times 10^{-17}$ rad^2/Hz (-81 dB rad^2/Hz) on the plot.

6. From $b_{-2}f^{-2} = b_0$, found on the plot, it follows that $f_L \simeq 25$ kHz. Hence, $Q = \nu_0(2f_L) \simeq 2 \times 10^5$. This is the typically good value obtained for a room-temperature WG sapphire resonator.

7. It is seen on the plot that $b_{-3} \simeq 7.9 \times 10^{-6}$ rad^2/Hz (-51 dB rad^2/Hz).

8. In a spectrum of type 2A, the corner frequency f_c of the sustaining amplifier is the frequency at which the oscillator noise changes slope from f^{-3} to f^{-2}. Thus, $f_c \simeq 850$ Hz. The flicker noise of the sustaining amplifier is $(b_{-1})_{ampli}f^{-1}$ with $(b_{-1})_{ampli} = b_0 f_c$. Thus $(b_{-1})_{ampli} \simeq 10^{-14}$ rad^2/Hz (-140 dB rad^2/Hz).

9. The white and flicker frequency noise, transformed into the Allan variance (Table 1.4), is

$$\sigma_y^2(\tau) = \frac{h_0}{2\tau} + 2\ln 2\, h_{-1}$$

$$= \frac{b_{-2}}{\nu_0^2}\frac{1}{2\tau} + 2\ln 2\, \frac{b_{-3}}{\nu_0^2}$$

$$\simeq \frac{7.9 \times 10^{-9}}{2 \times (10 \times 10^9)^2}\frac{1}{\tau} + 1.39 \times \frac{7.9 \times 10^{-6}}{(10 \times 10^9)^2};$$

thus

$$\sigma_y^2(\tau) \simeq \frac{4 \times 10^{-29}}{\tau} + 1.1 \times 10^{-25}$$

and so

$$\sigma_y(\tau) \asymp \frac{6.3 \times 10^{-15}}{\sqrt{\tau}} + 3.3 \times 10^{-13}.$$

10. The bump at $f = 40$ kHz is ascribed to the feedback control of the noise degeneration circuit.

UWA liquid-N whispering-gallery 9 GHz oscillator

When cooled to liquid-N temperature (77 K), a sapphire resonator improves significantly in all the relevant parameters. The quality factor increases from 2×10^5 at room

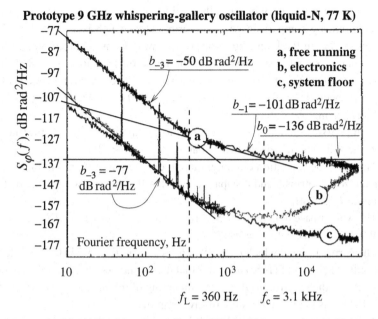

Figure 6.15 Liquid-N whispering-gallery 9 GHz oscillator tested at the University of Western Australia [108]. Used with the permission of the IEEE, 2007. The interpretation, comments, and any mistakes are those of the present author.

temperature to 3×10^7 (unloaded); the thermal coefficient of the resonant frequency decreases from -7×10^{-5}/K to -1×10^{-5}/K; and the thermal conductivity increases from 24 W m^{-1}K^{-1} to 960 W m^{-1}K^{-1}, which ensures temperature uniformity and in turn reduces the mechanical stress. The two reference books on these types of oscillator and resonator are [67] and [59].

Figure 6.15 shows the phase noise of a prototype 9 GHz oscillator built around a liquid-N whispering-gallery resonator, tested at the University of Western Australia [108]. This oscillator includes a noise-degeneration amplifier based on the ideas discussed in subsection 2.5.3. The phase-noise measurement of this type of oscillator is difficult, and for this reason [108] gives the result as a set of fragments of information instead of as a single phase-noise spectrum. Thus, we will analyze only the oscillator with the noise-degeneration loop disabled. The phase-noise spectrum is curve **a** of Fig. 6.15. This plot is fitted by the polynomial $\sum_{i=-3}^{0} b_i f^i$, with

b_0	-136 dB	2.5×10^{-14}	rad^2/Hz
b_{-1}	-101 dB	8×10^{-13}	rad^2/Hz
b_{-2}		(not visible)	
b_{-3}	-50 dB	1×10^{-5}	rad^2/Hz

We interpret the noise spectrum as follows.

1. The f^{-1} and f^{-3} noise types show up clearly, while there is no f^{-2} noise. Hence, the spectrum is of type 2 ($f_L < f_c$).

2. The white noise is only partially visible. This occurs because the oscillator is measured by comparison with a discriminator, which is a resonator of the same type as that in the loop (Fig. 5 of [108]). Away from the discriminator bandwidth the output signal loses power; thus the phase detector loses gain and the voltage noise at the detector output decreases. This leads to the wrong conclusion that the noise spectral density still decreases beyond 20–30 kHz. Trusting the theory, we find b_0 using two criteria:
   - the 3 dB difference between the straight-line approximation and the true spectrum;
   - the value in the narrow, nearly flat, region at $f \approx 10$ kHz.

3. The noise-degeneration circuit, although disabled, is still present in the circuit and leads to an intrinsic and dissipative loss. For this reason, we assume that the noise figure F is 5 dB.

4. The white phase noise $b_0 = 2.5 \times 10^{-14}$ rad^2/Hz. Thus $P_0 = FkT/b_0 = 5 \times 10^{-7}$ W (-33 dB m), with $F = 5$ dB.

5. The corner point at which $b_{-1}f^{-1} = b_0$ occurs at $f_c = 3.1$ kHz. The scheme of the oscillator (Fig. 3 of [108]) indicates that there is no output buffer. Hence the $b_{-1}f^{-1}$ region of the spectrum is due to the flickering of the sustaining amplifier. The value -101 dB is plausible for a microwave amplifier.

6. According to the scheme published, the oscillator is followed by two isolators and no buffer. With this configuration, isolation would probably be insufficient for practical applications. Yet this can be tolerated in a laboratory environment, with the oscillator connected to a stable load.

7. The frequency at which $b_{-3}f^{-3} = b_{-1}f^{-1}$, is $f_L = 360$ Hz. In the type-2 spectrum we have no means of saying whether this corner point is the true Leeson frequency rather than an effect of the fluctuating frequency of the resonator. Better a priori knowledge of the loaded quality factor is necessary, which we do not have. However, in this case we can find the answer somewhere else. In fact the f^{-3} noise improves significantly after introduction of the noise-degeneration circuit, since the resonator $1/f^3$ noise is lower than that of plot **a** in Fig. 6.15. Consequently, $f_L = 360$ Hz must be the true Leeson frequency.

8. With $f_L = 360$ Hz, we find $Q = \nu_0/f_L = 1.25 \times 10^7$. This is a reasonable value for the loaded quality factor of this type of resonator.

9. There is remarkable agreement between the two estimates of f_L, that based on the crossing point of the straight lines and that based on the 3 dB difference between the straight lines and the measured spectrum.

10. The frequency flicker transformed into the Allan variance (Table 1.4) is

$$\sigma_y^2(\tau) = 2\ln 2 \, h_{-1} = 2\ln 2 \, \frac{b_{-3}}{\nu_0^2} = 1.39 \times \frac{1 \times 10^{-5}}{(9 \times 10^9)^2} \, ;$$

thus

$$\sigma_y^2(\tau) \simeq 1.7 \times 10^{-25} \tag{6.11}$$

and so

$$\sigma_y(\tau) \simeq 4.2 \times 10^{-13} \, .$$

6.7   Optoelectronic oscillators

NIST 10 GHz opto-electronic oscillator (OEO)

The unique feature of the optoelectronic oscillator (OEO), developed at the Jet Propulsion Laboratory (JPL) [110] in the mid 1990s, is that the frequency-selective element is an optical delay line instead of a traditional resonator. The basic structure consists of a loop in which the microwave sinusoid modulates a laser beam. After traveling through a long optical fiber, the microwave signal is detected by a high-speed photodetector and fed back into the intensity modulator after amplification.

We analyze the noise of a prototype tested at the US National Institute of Standards and Technology (NIST) and published in the article [78]. The reason for choosing this oscillator is that the published article, as compared with other references, discloses more technical information specific to the prototype. The phase-noise spectrum, shown in Fig. 6.16, is fitted by the polynomial $\sum_{i=-4}^{0} b_i f^i$, with

$S_{\varphi\,min}$	-137 dB	2×10^{-14}	rad^2/Hz
b_{-1}		(not visible)	
b_{-2}	-42 dB	6.3×10^{-5}	rad^2/Hz
b_{-3}	-10 dB	0.1	rad^2/Hz
b_{-4}	$+2$ dB	1.6	rad^2/Hz

We interpret the spectrum as follows.

1. In the case of an OEO, the "sustaining amplifier" also includes the optical modulator and the photodetector. For short, the subscript 'ampli' will be used for notational consistency, instead of making explicit reference to the above chain of devices. The structure of this chain will be introduced when needed.

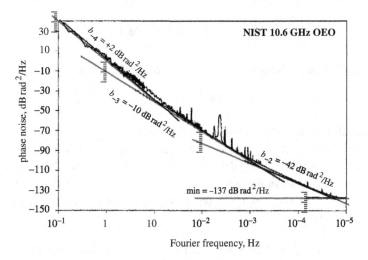

Figure 6.16 Phase noise of the NIST OEO prototype, employing 1.2 km optical fiber as a delay line. The spectrum shown is Fig. 7 of [78], used with the permission of the IEEE, 2007. The interpretation, comments, and any mistakes are those of the present author.

2. From [78], the length of the optical fiber is $L = 1.2$ km. With a typical refraction index of 1.468, the expected delay is $\tau_d = 5.88$ μs. Thus the base frequency must be $1/\tau_d = 170$ kHz.

3. From the phase-noise transfer function H(s), (5.69) (also shown in Fig. 5.14), we expect a minimum at $f \simeq 1/(2\tau_d) \approx 85$ kHz. At this minimum, it holds that $S_\varphi(f) = \frac{1}{4}S_\psi(f)$.

4. The sequence of peaks and minima is not visible in Fig. 6.16 because the frequency axis is truncated at 100 kHz. Only the minimum at 85 kHz is visible.

5. The value tabulated as $S_{\varphi\,\mathrm{min}} = 2 \times 10^{-14}$ rad²/Hz (-137 dB rad²/Hz) is not the floor b_0 of the oscillator. It is the first minimum predicted by (5.69).

6. It is seen in Fig. 6.16 that the corner frequency at which the $1/f^2$ noise turns into $1/f^3$ noise occurs at 1.6 kHz. This corresponds to the corner frequency f_c of the electronics, amplifier, and photodetector, where the noise turns from f^0 to $1/f$.

7. The minimum at 85 kHz is in the white-noise region of the electronics, where the corner frequency f_c is far enough away for the $1/f$ noise to be negligible. Thus, using $S_\varphi(f) = \frac{1}{4}S_\psi(f)$ we get

$$(b_0)_{\mathrm{ampli}} = 8 \times 10^{-14} \text{ rad}^2/\text{Hz} \qquad (-131 \text{ dB rad}^2/\text{Hz}) .$$

8. Comparing $(b_0)_{\mathrm{ampli}} = 8 \times 10^{-14}$ with $b_0 = FkT/P_0$ and using the microwave power $P_0 = 2 \times 10^{-7}$ W (-37 dB m, as published) we find

$$F = 4 \qquad (6 \text{ dB}) .$$

This value includes thermal noise, the white noise of the amplifier, and the photodetector's shot noise.

9. Equation (5.69) gives an expression for the phase-noise transfer function $|H(j2\pi f)|^2|$. Fitting the phase-noise spectrum of Fig. 6.16 to the model

$$S_\varphi(f) = |H(j2\pi f)|^2 \left(b_0 + \frac{b_{-1}}{f}\right) \tag{6.12}$$

and excluding the frequency random walk (b_{-4}/f^4), we find

$$(b_{-1})_{\mathrm{ampli}} = 1.6 \times 10^{-10} \text{ rad}^2/\text{Hz} \qquad (-98 \text{ dB rad}^2/\text{Hz}) .$$

This value includes the amplifier and the photodetector.

10. Feeding into $|H(j2\pi f)|^2$ the noise parameters b_0 and b_{-1}, which we have identified above, plus the filter group delay $\tau_f \simeq Q/(\pi\nu_0)$ and the fiber delay τ_d, we find the noise model shown in Fig. 6.17.

11. Figure 6 of [78], not shown here, reports the phase noise of the RF chain, suggesting the parameters

$$(b_{-1})_{\mathrm{ampli}}^{\mathrm{Fig.6}} = 1.26 \times 10^{-11} \text{ rad}^2/\text{Hz} \qquad (-109 \text{ dB rad}^2/\text{Hz}),$$

$$(b_0)_{\mathrm{ampli}}^{\mathrm{Fig.6}} = 1.26 \times 10^{-14} \text{ rad}^2/\text{Hz} \qquad (-139 \text{ dB rad}^2/\text{Hz})$$

$S_\varphi(f)$, dB rad²/Hz

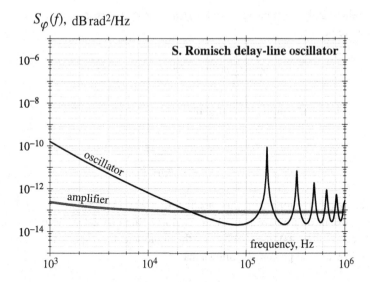

Figure 6.17 Phase-noise model of the the NIST OEO prototype (Fig. 6.16) excluding the frequency random walk. The parameters are as follows: $\tau = 5.9 \times 10^{-6}$ s, $m = 62\,300$, $\nu_m = 10.6$ GHz, $Q = 8300$, $b_0 = -131$ dB, $b_1 = -98$ dB.

measured at 10^{-7} W (-40 dB m), which scale to

$$(b_{-1})^{\text{Fig.6}}_{\text{ampli}} = 1.26 \times 10^{-11} \text{ rad}^2/\text{Hz} \qquad (-109 \text{ dB rad}^2/\text{Hz}),$$

$$(b_0)^{\text{Fig.6}}_{\text{ampli}} = 6.3 \times 10^{-15} \text{ rad}^2/\text{Hz} \qquad (-142 \text{ dB rad}^2/\text{Hz})$$

at the actual operating power, 2×10^{-7} W (-37 dB m). Notice that the phase noise scales with power while the flicker phase noise does not, as explained in Section 2.4.

12. There is a systematic inconsistency of 11 dB between our values, calculated from the oscillator phase noise, and the values obtained from Fig. 6 of [78]. The inconsistency is the same for flicker and white noise.

13. Having detected an inconsistency, we have to decide what to trust. So, we will work from the phase-noise spectrum of Fig. 6.16, discarding Fig. 6 of [78], for the following reasons.
    - Looking at Fig. 6 of [78], at the corner point the plotted spectrum differs from the asymptotic approximation by 7 dB instead of 3 dB.
    - The noise spectrum is said to have been measured at -40 dB m input power. The floor of -139 dB rad²/Hz, found on the figure, is incompatible with the carrier power. At -40 dB m input power, and at room temperature, the law $b_0 = FkT/P_0$ gives a minimum floor of -134 rad²/Hz for a noise-free amplifier.
    - The phase-noise spectrum of Fig. 6.16 is central in [78], while Fig. 6 is not. To this extent, the inconsistency does not even disturb one's reading of this article.

14. Introducing the responsivity $\rho = q\eta/(h\nu_l)$ into the expression (2.80) for the microwave power,

$$P_0 = \frac{1}{2} m^2 R_0 \rho^2 P_l^2 \qquad \text{(from (2.80))},$$

(the averages are implied) we find the modulation index

$$m = \sqrt{\frac{2P_0}{R_0 \rho^2 P_l^2}}.$$

The practical values for this oscillator, given in [78], are $P_0 = 2 \times 10^{-7}$ W (-37 dB m microwave power), $R_0 = 50\ \Omega$ (the characteristic impedance for all microwave circuits), $\rho = 0.17$ A/W (the photodetector responsivity), and $P_l = 1.75$ mW (the optical power). It follows that

$$m \simeq 0.3 \qquad \text{(modulation index)}.$$

15. The photodetector responsivity, $\rho = 0.17$ A/W and also the modulation index $m \simeq 0.3$ are unusually low. This explains the low microwave power.

16. The white and flicker frequency noise, transformed into the Allan variance (Table 1.4), is

$$\sigma_y^2(\tau) = \frac{h_0}{2\tau} \simeq \frac{6.3 \times 10^{-5}}{2 \times (10.6 \times 10^9)^2} \frac{1}{\tau} = \frac{1.23 \times 10{-11}}{\tau} \qquad \text{(white)},$$

$$\sigma_y^2(\tau) = 2 \ln 2\, h_{-1} \simeq \frac{1.39 \times 0.1}{(10.6 \times 10^9)^2} = 1.23 \times 10^{-21} \qquad \text{(flicker)},$$

$$\sigma_y^2(\tau) = \frac{(2\pi)^2}{6} h_{-2} \tau \simeq \frac{6.58 \times 1.6}{2 \times (10.6 \times 10^9)^2} \tau = 9.3 \times 10^{-20}\, \tau, \qquad \text{(r. walk)}.$$

Thus

$$\sigma_y^2(\tau) \simeq \frac{1.23 \times 10 - 11}{\tau} + 1.23 \times 10^{-21} + 9.3 \times 10^{-20}\, \tau$$

and so

$$\sigma_y(\tau) \asymp \frac{5.3 \times 10^{-13}}{\sqrt{\tau}} + 3.5 \times 10^{-11} + 3.05 \times 10^{-10}\, \sqrt{\tau}.$$

OEwaves Tidalwave (10 GHz OEO)

Apart from laboratory prototypes, OEwaves is up to the present time the only manufacturer of OEOs. When studying oscillator noise, the Tidalwave is a challenging example because the general experience earned with traditional oscillators is now of scarce utility.

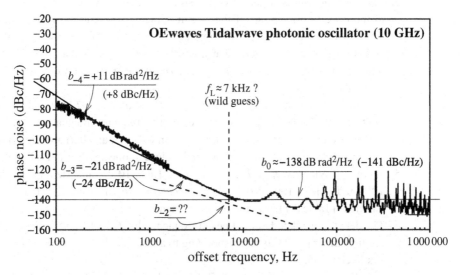

Figure 6.18 Phase-noise spectrum of the Tidalwave photonic oscillator (patents 5 777 778, 6 928 091, and 7 184 451). Courtesy of OEwaves Inc., 2008. The interpretation, comments, and any mistakes are those of the present author.

The phase-noise spectrum (Fig. 6.18) is fitted by the polynomial $\sum_{i=-4}^{0} b_i f^i$, with

b_0	-138 dB	1.6×10^{-14}	rad^2/Hz
b_{-1}		(not visible)	
b_{-2}		(not visible)	
b_{-3}	-21 dB	7.9×10^{-3}	rad^2/Hz
b_{-4}	$+11$ dB	1.26×10^{1}	rad^2/Hz

We interpret the noise spectrum as follows.

1. In a delay-line oscillator with delay τ we expect (see Chapter 5) a "clean" comb of spectral lines at $f_l = l/\tau_d$, integer $l \geq 1$. This structure is not present in the spectrum of Fig. 6.18. Instead, there is a series of smaller bumps and spectral lines. This indicates that the Tidalwave is not a simple delay-line oscillator and that it contains some additional circuits that reduce or almost eliminate the spectral lines at $f_\mu = \mu/\tau_d$, integer μ. The remedy suggested by the literature is the dual-loop oscillator, in which a second delay line is used to significantly reduce the largest peaks. The references [111, 29] provide information that is interesting yet insufficient by far for identifying the internal details of this specific oscillator.

2. On the right-hand side of the spectrum, i.e., in the white-noise region, a horizontal b_0 line gives only a basic approximation to the spectrum. This is an inherent feature of the delay-line oscillator. Discarding the peaks, we take $b_0 \approx 1.6 \times 10^{-14}$ rad^2/Hz (-138 dB rad^2/Hz).

3. The equivalent-noise figure used to estimate the microwave power must be referred to the lowest-power point of the circuit, which is the output of the photodetector. The noise includes at least the intensity fluctuation of the laser, the shot noise, and the noise of the microwave amplifier. Let us guess that $F = 6$ dB. Consequently, the microwave power $P_0 = FkT/b_0$ at the phodetector output is $P_0 = 1$ µW $(-30$ dB m).

4. One may think that the transition $f^{-3} \to f^0$ around $f = 7$ kHz is the signature of the frequency flicker of the optical fiber, cf. the frequency fluctuation of the resonator in the case of the Wenzel oscillator (Section 6.4). Yet an alternative interpretation is possible here because f_c and f_L might be close to one another and fall in this region.

5. A corner frequency $f_c \approx 7$ kHz with $b_0 \approx 1.6 \times 10^{-14}$ rad^2/Hz requires that $b_{-1} = 10^{-10}$ rad^2/Hz $(-100$ dB rad^2/Hz). The latter value can be ascribed to the microwave amplifier (Table 2.1) or to the optical system.

6. The delay τ_d is related to the Leeson frequency by $f_L = 1/(2\pi\tau_d)$. Thus $f_L \approx 7$ kHz requires that $\tau_d \approx 23$ µs, corresponding to a fiber length $l = c\tau_d/n \approx 4.7$ km, which is a likely value.

7. The $f^{-3} \to f^0$ transition at $f \approx 7$ kHz can be ascribed either to the fact that $f_c \approx f_L$ or to the flickering of the delay. The analysis of the spectrum is not sufficient for us to be able to say more.

8. The frequency flicker and random walk, transformed into the Allan variance (Table 1.4), is

$$\sigma_y^2(\tau) = 2 \ln 2\, h_{-1} + \frac{4\pi^2}{6} h_{-2}\tau$$

$$= 2 \ln 2 \frac{b_{-3}}{v_0^2} + \frac{4\pi^2}{6} \frac{b_{-4}}{v_0^2} \tau$$

$$= 1.39 \times \frac{7.9 \times 10^{-3}}{(10^{10})^2} + \frac{4\pi^2}{6} \times \frac{12.6}{(10^{10})^2} \tau \;;$$

thus

$$\sigma_y^2(\tau) \simeq 1.1 \times 10^{-22} + 8.3 \times 10^{-18}\, \tau$$

and so

$$\sigma_y(\tau) \asymp 1.05 \times 10^{-11} + 9 \times 10^{-9}\, \sqrt{\tau}\,.$$

Exercises

6.1 Find on the web the phase-noise spectra of some oscillators analyzed in this chapter. Repeat the analysis yourself and compare the results. If possible, fit the spectra with a power law using a set square and rule. Feel free to use Chapters 1 to 3 but refrain from using this chapter until you have finished.

6.2 Hack as many microwave oscillators as you can, chosen in your domain of interest.

The reader will begin to realize that there is very little information on the web and that the literature does not help much. Trade secrets are everywhere. Quartz and microwave oscillators are often based on old ideas and new technology. Many delay-line oscillators are found in optics because most lasers are actually oscillators of this type. In lasers, it is generally difficult or impossible to separate the resonator from the amplifier.

Dear reader, thank you for getting to this page. From now you are alone in the real world . . .

Appendix A   Laplace transforms

This appendix provides a very short summary of useful properties and formulae, focusing on the first-order low-pass system function (Fig. A.1) and on the second-order resonator system function (Fig. A.2). The reader is encouraged to refer to the numerous textbooks that are available, among which we prefer [66, 97].

The formulae for the Laplace transform of a function $f(t)$ and for the inverse transformation are

$$F(s) = \mathcal{L}\{f(t)\} = \int_0^\infty f(t)e^{-st}dt \qquad \text{(Laplace transform)} \qquad \text{(A.1)}$$

$$f(t) = \mathcal{L}^{-1}\{f(t)\} = \frac{1}{j2\pi} \int_{\sigma-j\infty}^{\sigma+j\infty} F(s)e^{st}ds \qquad \text{(inversion formula)} . \qquad \text{(A.2)}$$

The Laplace transform can be seen as an extension of the Fourier transform, on replacing $j\omega$ by $s = \sigma + j\omega$, and is particularly useful when $f(t)$ is a causal function, i.e. $f(t) = 0$ for $t < 0$. Actual time-domain systems are always causal. That being said, the relation between the Fourier and Laplace transforms is found as follows:

$$\mathcal{F}\{f(t)\} = \int_{-\infty}^\infty f(t)e^{-j\omega t}dt \qquad \text{(Fourier transform)} \qquad \text{(A.3)}$$

$$= \int_0^\infty f(t)e^{-j\omega t}dt \qquad \begin{array}{l}\text{(causal signal,}\\ f(t) = 0 \text{ for } t < 0)\end{array} \qquad \text{(A.4)}$$

$$= \mathcal{L}\{f(s)\}\big|_{s=j\omega} . \qquad \text{(A.5)}$$

Table A.1 gives properties of Laplace transforms and Table A.2 lists the transforms of various functions.

Table A.1 Properties of Laplace transforms

Property	$f(t)$	$F(s)$	Condition
linearity	$af(t)$ $f_1(t) + f_2(t)$	$aF(s)$ $F_1(s) + F_2(s)$	
convolution	$f_1(t) * f_2(t)$	$F_1(s) F_2(s)$	
derivative	$f'(t)$	$sF(s)$	$f(0_-) = 0$
integral	$\int f(t)\, dt$	$\dfrac{1}{s} F(s)$	
time stretch	$f(at)$	$\dfrac{1}{a} F\left(\dfrac{s}{a}\right)$	
frequency stretch	$\dfrac{1}{a} f\left(\dfrac{f}{a}\right)$	$F(as)$	
time translation	$f(t - a)$	$e^{-as} F(s)$	$a \geq 0$
frequency translation	$e^{-at} f(t)$	$F(s - a)$	

Table A.2 Useful Laplace transforms

Function	$f(t)$	$F(s)$	Condition
Dirac	$\delta(t)$	1	
Dirac (delayed)	$\delta(t - a)$	e^{-as}	$a \geq 0$
Heaviside	$U(t)$	$1/s$	
exponential	e^{-at}	$\dfrac{1}{s + a}$	$s > -a$
sinusoid	$\sin \omega t$	$\dfrac{\omega}{s^2 + \omega^2}$	
cosinusoid	$\cos \omega t$	$\dfrac{s}{s^2 + \omega^2}$	
damped sinusoid	$e^{-at} \sin \omega t$	$\dfrac{\omega}{(s + a)^2 + \omega^2}$	
damped cosinusoid	$e^{-at} \cos \omega t$	$\dfrac{s + a}{(s + a)^2 + \omega^2}$	

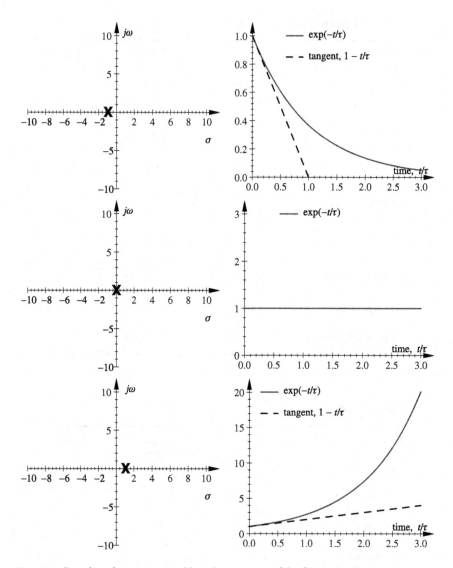

Figure A.1 Complex-plane pattern and impulse response of the first-order (low-pass) system function $F(s) = 1/(s + 1/\tau)$. Poles are represented on the complex plane as crosses.

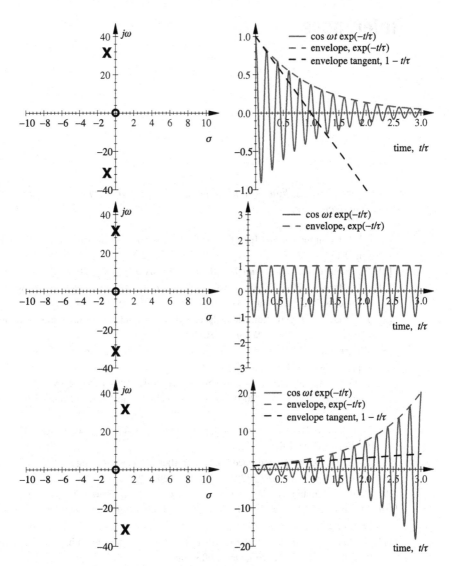

Figure A.2 Complex plane pattern and impulse response of the second-order (resonator) system function $F(s) = s/(s^2 + 2s/\tau + \omega_n^2)$. Poles and zeros are represented on the complex plane as crosses and circles, respectively.

References

Sections, subsections, figures, and tables in which a publication is mentioned are listed in order of occurrence.

[1] AML Communications Inc., Camarillo CA, http://www.amlj.com/. Table 2.2, 2.5.4, Fig. 2.17

[2] P. Ashburn. *SiGe Heterojunction Bipolar Transistors*. Wiley, 2003. 2.5.6

[3] J. A. Barnes and D. W. Allan. Variances based on data with dead time between the measurements. Technical note 1318, NIST, Boulder CO, 1990. 1.7.1

[4] J. Bernamont. Fluctuations de résistance dans un conducteur métallique de faible volume. *Comptes Rendus Acad. Sciences (Paris)*, 198: 1755–1758, 1934. 2.1.3

[5] J. Bernamont. Fluctuations in the resistance of thin films. *Proc. Phys. Soc. (London)*, 49 suppl.: 138–139, July 1937. 2.1.3

[6] E. D. Black. An introduction to Pound–Drever–Hall laser frequency stabilization. *Am. J. Phys.*, 69(1): 79–87, Jan. 2001. Also LIGO technical note LIGO-T980045-00D, May 1998. 3.4.3

[7] H. S. Black. Translation system. US Patent no. 1 686 792, issued 9 Oct. 1928 (concept of feedforward amplifier). 2.5.2

[8] H. S. Black. Stabilized feed-back amplifiers. *Electrical Eng.*, 53(1): 114–120, 1934. Reprinted in *Proc. IEEE*, 87(2): 379–385, Feb. 1999. 2.5.2

[9] R. B. Blackman and J. W. Tuckey. *The Measurement of Power Spectra*. Dover, 1959. 1.4.1

[10] M. Born and E. Wolf. *Principles of Optics*. Cambridge University Press, seventh edition, 1999. 5.2.1

[11] R. Boudot. *Oscillateurs micro-onde à haute pureté spectrale*. Ph.D. thesis, Université de Franche Comté, Besançon, Dec. 2006. Fig. 2.8

[12] R. Boudot, Y. Gruson, N. Bazin, E. Rubiola, and V. Giordano. Design and measurement of a low phase-noise X-band oscillator. *Electron. Lett.*, 42(16): 929–931, Aug. 2006. 2.5.6

[13] S. Bregni. *Synchronization of Digital Telecommunications Networks*. Wiley, 2002. 1.7.2

[14] O. E. Brigham. *The Fast Fourier Transform and its Applications*. Prentice-Hall, 1988. 1.4.3

[15] J. R. Burgoon and R. L. Wilson. SC-cut quartz oscillator offers improved performance. *Hewlett Packard J.*: 20–29, Mar. 1981. Fig. 6.8, 6.4, Table 6.1

[16] V. Candelier, P. Canzian, J. Lamboley, M. Brunet, and G. Santarelli. Ultra stable oscillators. In *Proc. Europ. Freq. Time Forum and Freq. Control Symp. Joint Meeting*, Tampa FL, May 2003, pp. 575–582. Table 6.1

[17] V. Candelier, J. Chauvin, C. Gellé, G. Marotel, M. Brunet, and R. Petit. Ultra stable oscillators. In *Proc. Europ. Freq. Time Forum*, Warszawa, Mar. 1998, pp. 345–351. Table 6.1

[18] S. Chang, A. G. Mann, A. N. Luiten, and D. G. Blair. Measurements of radiation pressure effect in cryogenic sapphire dielectric resonators. *Phys. Rev. Lett.*, 79(11): 2141–2144, Sept. 1997. 1.6.4

[19] C. J. Christensen and G. L. Pearson. Spontaneous resistance fluctuations in carbon microphones and other granular resistances. *Bell Syst. Techn. J.*, 15(2): 197–223, Apr. 1936. 2.1.3

[20] G. Cibiel, M. Régis, O. Llopis, A. Rennane, L. Bary, R. Plana, Y. Kersalé, and V. Giordano. Optimization of an ultra low-phase noise sapphire: SiGe HBT oscillator using nonlinear CAD. *IEEE Trans. Ultras. Ferroelec. and Freq. Contr.*, 51(1): 33–41, Jan. 2004. 2.5.6

[21] http://www.zakou.com and http://www.cmac.com/. 2.5.6, Table 6.1

[22] J. D. Cressler, editor. *Silicon Heterostructure Handbook*. CRC, 2006. 2.5.6

[23] J. D. Cressler and G. Niu. *Silicon–Germanium Heterojunction Bipolar Transistors*. Artech House, 2003. 2.5.6

[24] B. Cretin and J. C. Mollier. Design of a 3/2-order phase-locked loop for improved laser probe resolution. *IEEE Trans. Instrum. Meas.*, 34(4): 660–664, Dec. 1985. 6.6

[25] S. T. Dawkins, J. J. McFerran, and A. N. Luiten. Considerations on the measurement of the stability of oscillators with frequency counters. *IEEE Trans. Ultras. Ferroelec. and Freq. Contr.*, 54(5): 918–925, May 2007. 1.7

[26] R. W. P. Drever, J. L. Hall, F. V. Kowalski, J. Hough, G. M. Ford, A. J. Munley, and H. Ward. Laser phase and frequency stabilization using an optical resonator. *Appl. Phys. Lett.*, 31(2): 97–105, June 1983. Fig. 3.18, 3.4.3

[27] M. M. Driscoll and R. W. Weinert. Spectral performance of sapphire dielectric resonator-controlled oscillators operating in the 80 K to 275 K temperature range. In *Proc. Freq. Control Symp.*, San Francisco CA, May–June 1995, pp. 401–412. 2.5.1, 2.5.3

[28] B. C. Edwards, M. I. Buchwald, and R. I. Epstein. Development of the Los Alamos solid-state optical refrigerator. *Rev. Sci. Instrum.*, 69(5): 2050–2055, May 1998. 2.6.6

[29] D. Eliyahu and L. Maleki. Low phase noise and spurious level in multi-loop opto-electronic oscillators. In *Proc. Europ. Freq. Time Forum and Freq. Control Symp. Joint Meeting*, Tampa FL, May, 2003, pp. 405–410.

[30] R. I. Epstein, M. I. Buchwald, B. C. Edwards, T. R. Gosnell, and C. E. Mungan. Observation of a laser-induced fluorescent cooling of a solid. *Nature*, 377(6549): 500–503, Oct. 1995. 6.6

[31] J. K. A. Everard and M. A. Page-Jones. Ultra low noise microwave oscillators with low residual flicker noise. In *Proc. Int. Microw. Theory Tech. Symp.*, May 1995, pp. 693–696. 2.5.1

[32] W. Feller. *An Introduction to Probability Theory and Its Applications*, vol. 2. Wiley, second edition, 1971. 1.3

[33] H. T. Friis. Noise figure of radio receivers. *Proc. IRE*, 32: 419–422, July 1944. 2.2.2

[34] Z. Galani, M. J. Bianchini, R. C. Waterman, Jr., R. Dibiase, R. W. Laton, and J. Bradford Cole. Analysis and design of a single-resonator GaAs FET oscillator with noise degeneration. *IEEE Trans. Microw. Theory Tech.*, 32(12): 1556–1565, Dec. 1984. Fig. 3.20, 3.4.3, 4.5.3

[35] S. Galliou, F. Sthal, J.-J. Boy, and M. Mourey. Recent results on quartz crystal LD-cuts operating in oscillators. In *Proc. Ultrason. Ferroelec. Freq. Contr. Joint Conf.*, Montreal, Aug. 2004, pp. 475–477. 6.4, Table 6.1

[36] E. A. Gerber and A. Ballato. *Precision Frequency Control* (two volumes). Academic Press, 1985. 6.3

[37] L. J. Giacoletto. Study of p–n–p alloy junction transistor from dc through medium frequencies. *RCA Rev.*, 15: 506–562, Dec. 1954. 2.2.3

[38] M. L. Gorodetsky, A. A. Savchenkov, and V. S. Ilchenko. Ultimate Q of optical microsphere resonators. *Optics Lett.*, 21(7): 453–455, Apr. 1996. 5.2.2

[39] C. A. Greenhall. Spectral ambiguity of Allan variance. *IEEE Trans. Instrum. Meas.*, 47(3): 623–627, June 1998. 1.8

[40] I. S. Grudinin, A. B. Matsko, A. A. Savchenkov, D. Strekalov, V. S. Ilchenko, and L. Maleki. Ultra high Q crystalline microcavities. *Optics Comm.*, 265: 33–38, Sept. 2006. 5.2.2

[41] A. Gruhle and C. Mähner. Low $1/f$ noise SiGe HBTs with application to low phase noise microwave oscillators. *Electron. Lett.*, 33(24): 2050–2052, Nov. 1997. 2.5.6

[42] N. Gufflet, R. Bourquin, and J.-J. Boy. Isochronism defect for various doubly rotated cut quartz resonators. *IEEE Trans. Ultras. Ferroelec. and Freq. Contr.*, 49(4): 514–518, Apr. 2002. 1.6.4, 6.4

[43] E. A. Guillemin. *Synthesis of Passive Networks*. Wiley, 1957. 4.8

[44] H. K. Gummel and H. C. Poon. An integral charge control model of bipolar transistors. *Bell Syst. Techn. J.*, 49: 827, May/June 1970. 2.2.3

[45] D. Halford, A. E. Wainwright, and J. A. Barnes. Flicker noise of phase in RF amplifiers: characterization, cause, and cure. In *Proc. Freq. Control Symp.*, Apr. 1968, pp. 340–341. Abstract only is published. 2.3.2

[46] A. Hati, D. Howe, D. Walker, and F. Walls. Noise figure vs. PM noise measurements: a study at microwave frequencies. In *Proc. Europ. Freq. Time Forum and Freq. Control Symp. Joint Meeting*, Tampa FL, May 2003. 2.3.2

[47] H. Hellwig (chair.). *IEEE Standard Definitions of Physical Quantities for Fundamental Frequency and Time Metrology*. IEEE Standard 1139–1988. 1.6.1

[48] W. R. Hewlett. *A New Type Resistance–Capacity Oscillator*. Graduate thesis, Leland Stanford Junior University, Palo Alto CA, June 1939. 4.3.2

[49] F. N. Hooge. $1/f$ noise is no surface effect. *Phys. Lett. A*, 29: 139–140, 1969. 2.1.3, 2.5.5

[50] C. H. Horn. A carrier suppression technique for measuring S/N and carrier/sideband ratios greater than 120 dB. In *Proc. Freq. Control Symp.*, Fort Monmouth NJ, May 1969, pp. 223–235. 1.9

[51] http://www.hparchive.com/Manuals/. 6.4

[52] V. S. Ilchenko and A. B. Matsko. Optical resonators with whispering-gallery modes – Part II: Applications. *J. Selected Topics Quantum Elec.*, 12(1): 15–32, Jan.–Feb. 2006. 5.2.2

[53] V. S. Ilchenko, X. S. Yao, and L. Maleki. Pigtailing the high-Q microsphere cavity: a simple fiber coupler for optical whispering-gallery modes. *Optics Lett.*, 24(11): 723–725, June 1999. 5.2.2

[54] E. N. Ivanov, J. G. Hartnett, and M. E. Tobar. Cryogenic microwave amplifiers for precision measurements. *IEEE Trans. Ultras. Ferroelec. and Freq. Contr.*, 47(6): 1273–1274, Nov. 2000. Fig. 2.9, 2.3.3

[55] E. N. Ivanov, M. E. Tobar, and R. A. Woode. Application of interferometric signal processing to phase-noise reduction in microwave oscillators. *IEEE Trans. Microw. Theory Tech.*, 46(10): 1537–1545, Oct. 1998. 2.5.3

[56] G. M. Jenkins and D. G. Watts. *Spectral Analysis and its Applications*. Holden Day, 1968. 1.4.1, 6.1.2

[57] J. B. Johnson. The Schottky effect in low frequency circuits. *Phys. Rev.*, 26: 71–85, July 1925. 2.1.3

[58] J. B. Johnson. Thermal agitation of electricity in conductors. *Phys. Rev.*, 32: 97–109, July 1928. 2.1.12

[59] D. Kajfez and P. Guillon, editors. *Dielectric Resonators*. Noble, second edition, 1998. Previously published by Artech House, 1986, and Vector Fields, 1990. 6.6, 6.6

[60] R. K. Karlquist. A new type of balanced-bridge controlled oscillator. *IEEE Trans. Ultras. Ferroelec. and Freq. Contr.*, 47(2): 390–403, Mar. 2000. Fig. 6.9, 6.4, 6.4. Table 6.1

[61] H. G. Kimball, editor. *Handbook of Selection and Use of Precise Frequency and Time Systems*. ITU, 1997. 1.6

[62] V. F. Kroupa. Theory of $1/f$ noise – a new approach. *Phys. Lett. A*, 336: 126–132, Jan. 2005. 6.3

[63] D. B. Leeson. A simple model of feed back oscillator noise spectrum. *Proc. IEEE*, 54: 329–330, Feb. 1966. 3.2

[64] D. B. Leeson and G. F. Johnson. Short-term stability for a Doppler radar: requirements, measurements, and techniques. *Proc. IEEE*, 54: 244, Feb. 1996 (document), 3.2.

[65] R. Leier. SiGe silences YIG oscillator phase noise. Electronic Design, Online ID no. 11902, Jan. 2006. 2.5.6

[66] W. R. LePage. *Complex Variables and Laplace Transform for Engineers*. McGraw Hill, 1961. Reprinted by Dover Books. Appendix A

[67] A. N. Luiten, editor. *Frequency Measurement and Control*. Topics in Applied Physics. Springer, 2001. 6.6, 6.6

[68] A. B. Matsko and V. S. Ilchenko. Optical resonators with whispering-gallery modes – Part I: Basics. *J. Selected Topics Quantum Elec.*, 12(1): 3–14, Jan.–Feb. 2006. 5.2.2

[69] A. L. McWhorter. $1/f$ noise and germanium surface properties. In R. H. Kingston, editor, *Semiconductor Surface Physics*, pp. 207–228. University of Pennsylvania Press, 1957. 2.1.3, 2.5.5

[70] G. K. Montress, T. E. Parker, and M. J. Loboda. Residual phase noise of VHF, UHF, and microwave components. *IEEE Trans. Ultras. Ferroelec. and Freq. Contr.*, 41(5): 664–679, Sept. 1994. 2.3.4

[71] C. W. Nelson, A. Hati, D. A. Howe, and W. Zhou. Microwave optoelectronic oscillator with optical gain. In *Proc. Intl. Freq. Control Symp. and Europ. Freq. Time Forum Joint Meeting*, Geneva, May–June 2007, pp. 1014–1019. 2.6.2

[72] H. Nyquist. Thermal agitation of electric charge in conductors. *Phys. Rev.*, 32: 110–113, July 1928. 2.1.1

[73] http://www.oscilloquartz.com/. Table 6.1

[74] A. Papoulis. *Probability, Random Variables and Stochastic Processes*. McGraw Hill, third edition, 1992. 1.3

[75] D. B. Percival and A. T. Walden. *Spectral Analysis for Physical Applications*. Cambridge University Press, 1998. 1.4.1, 6.1.2

[76] N. Pothecary. *Feedforward Linear Power Amplifiers*. Artech House, 1999. 2.5.2

[77] R. V. Pound. Electronic frequency stabilization of microwave oscillators. *Rev. Sci. Instrum.*, 17(11): 490–505, Nov. 1946. Fig. 3.17, 3.4.3, 4.5.3

[78] S. Römisch, J. Kitching, E. F. Pikal, L. Hollberg, and F. L. Walls. Performance evaluation of an optoelectronic oscillator. *IEEE Trans. Ultras. Ferroelec. and Freq. Contr.*, 47(5): 1159–1165, Sept. 2000. 6.7, Fig. 6.16

[79] E. Rubiola. The measurement of AM noise of oscillators. Website arXiv.org, document arXiv:physics/0512082, Dec. 2005. 1.6.4, 3.1.1

[80] E. Rubiola. On the measurement of frequency and of its sample variance with high-resolution counters. *Rev. Sci. Instrum.*, 76(5), May 2005. Websites arxiv.org and rubiola.org, document arXiv:physics/0411227, Dec. 2004. 1.7

[81] E. Rubiola and R. Boudot. The effect of AM noise on correlation phase noise measurements. Website arxiv.org, document arXiv:physics/0609147, Sept. 2006. 1.9.2

[82] E. Rubiola and V. Giordano. Advanced interferometric phase and amplitude noise measurements. *Rev. Sci. Instrum.*, 73(6): 2445–2457, June 2002. Website arxiv.org, document arXiv:physics/0503015v1. 2.5.3

[83] E. Rubiola and V. Giordano. On the 1/f frequency noise in ultra-stable quartz oscillators. *IEEE Trans. Ultras. Ferroelec. and Freq. Contr.*, 54(1): 15–22, Jan. 2007. Fig. 2.7, Fig. 3.11, Fig. 6.2, 6.3, Table 6.1

[84] E. Rubiola, J. Groslambert, M. Brunet, and V. Giordano. Flicker noise measurement of HF quartz resonators. *IEEE Trans. Ultras. Ferroelec. and Freq. Contr.*, 47(2): 361–368, Mar. 2000. 6.3

[85] E. Rubiola, Y. Gruson, and V. Giordano. On the flicker noise of circulators for ultra-stable oscillators. *IEEE Trans. Ultras. Ferroelec. and Freq. Contr.*, 51(8): 957–963, Aug. 2004. Fig. 3.17, Fig. 3.20

[86] E. Rubiola and F. Lardet-Vieudrin. Low flicker-noise amplifier for 50 Ω sources. *Rev. Sci. Instrum.*, 75(5): 1323–1326, May 2004. Free preprint available on the website arxiv.org, document arXiv:physics/0503012v1, March 2005. 2.3.4

[87] E. Rubiola, E. Salik, N. Yu, and L. Maleki. Phase noise measurement of low-power signals. In *Proc. Europ. Freq. Time Forum and Freq. Control Symp. Joint Meeting*, Montreal, Aug. 2004, pp. 292–297. Fig. 2.6

[88] E. Rubiola, E. Salik, N. Yu, and L. Maleki. Flicker noise in high-speed p–i–n photodiodes. *IEEE Trans. Microw. Theory Tech.*, 54(2): 816–820, Feb. 2006. Preprint available on arxiv.org, document arXiv:physics/0503022v1, March 2005. 2.6.2

[89] J. Rutman. Characterization of phase and frequency instabilities in precision frequency sources: fifteen years of progress. *Proc. IEEE*, 66(9): 1048–1075, Sept. 1978. 1.6

[90] B. E. A. Saleh and M. C. Teich. *Fundamentals of Photonics*. Wiley, 1991. 5.2.1

[91] A. S. Sedra and K. C. Smith. *Microelectronic Circuits*. Oxford University Press, fifth edition, 2004. 3.1

[92] H. Seidel. A microwave feedforward experiment. *Bell Syst. Techn. J.*, 50(9): 2879–2916, Nov. 1971.

[93] W. Shieh, X. S. Yao, L. Maleki, and G. Lutes. Phase-noise characterization of optoelectronic components by carrier suppression techniques. In *Proc. Optical Fiber Comm. Conf.*, San José CA, May 1998, pp. 263–264. 2.6.2

[94] A. E. Siegman. *Lasers*. University Science Books, 1986. 5.2.1

[95] J. Sikula, S. Hashiguchi, M. Ohki, and M. Tacano. Some considerations for the construction of low-noise amplifiers in very low frequency region. In J. Sikula and M. Levinsthein, editors, *Advanced Experimental Methods in Noise Research on Nanoscale Electronic Devices*, pp. 237–244. Kluwer, 2004. 2.3.4

[96] J. J. Snyder. Algorithm for fast digital analysis of interference fringes. *Appl. Opt.*, 19(4): 1223–1225, Apr. 1980. 1.7.2

[97] M. R. Spiegel. *Schaum's Outline of Laplace Transforms*. McGraw Hill, 1965. Appendix A

[98] P. Sulzer. High stability bridge balancing oscillator. *Proc. IRE*, 43(6): 701–707, June 1955. Fig. 3.19, 3.4.3

[99] K. K. Thladhar, G. Jenni, and J. Aubry. Improved BVA resonator – oscillator performances and frequency jumps. In *Proc. Europ. Freq. Time Forum*, Neuchâtel, Mar. 1997, pp. 273–280. Table 6.1

[100] A. van der Ziel. *Noise in Solid State Devices and Circuits*. Wiley, 1986. 2.2.3, 2.2.3

[101] M. E. Van Valkenburg. *Modern Network Synthesis*. Wiley, 1960. 4.8

[102] J. Vanier and C. Audoin. *The Quantum Physics of Atomic Frequency Standards*. Adam Hilger, 1989. 1.6

[103] J. R. Vig (chair.). *IEEE Standard Definitions of Physical Quantities for Fundamental Frequency and Time Metrology – Random Instabilities (IEEE Standards 1139–1999)*. IEEE, 1999. 1.6, 1.6.1, 1.6.1

[104] F. L. Walls. The quest to understand and reduce $1/f$ noise in amplifiers and BAW quartz oscillators. In *Proc. Europ. Freq. Time Forum*, Besançon, Mar. 1995, pp. 227–243. 6.3

[105] F. L. Walls, E. S. Ferre-Pikal, and S. R. Jefferts. Origin of $1/f$ PM and AM noise in bipolar junction transistor amplifiers. *IEEE Trans. Ultras. Ferroelec. and Freq. Contr.*, 44(2): 326–334, Mar. 1997. 2.3.2

[106] L. Weinberg. *Network Analysis and Synthesis*. McGraw Hill, 1962. 4.8

[107] http://www.wenzel.com. Table 6.1

[108] R. A. Woode, M. E. Tobar, E. N. Ivanov, and D. Blair. An ultra low noise microwave oscillator based on high Q liquid nitrogen cooled sapphire resonator. *IEEE Trans. Ultras. Ferroelec. and Freq. Contr.*, 43(5): 936–941, Sept. 1996. Fig. 6.15, 6.6

[109] X. S. Yao, L. Davis, and L. Maleki. Coupled optoelectronic oscillators for generating both RF signal and optical pulses. *J. Lightwave Technol.*, 18(1): 73–78, Jan. 2000. 4.4.1

[110] X. S. Yao and L. Maleki. Optoelectronic microwave oscillator. *J. Opt. Soc. Am. B – Opt. Phys.*, 13(8): 1725–1735, Aug. 1996. 5.1, 5.8, 6.7

[111] X. S. Yao and L. Maleki. Multiloop optoelectronic oscillator. *J. Quantum Electron.*, 36(1): 79–84, Jan. 2000. 6.6

Index